Project Managem
Small Business

How to Deliver Successful, Profitable Projects on Time with Your Small Business Clients

Dana J Goulston, PMP

and

Karl W. Palachuk

Published by

Great Little Book Publishing

Sacramento, CA

www.GreatLittleBook.com

Great Little Book Publishing

Sacramento, CA

Project Management in Small Business: How to Deliver Successful, Profitable Projects on Time with Your Small Business Clients

Parts of this book are derived from or previously appeared in *The Super-Good Project Planner for Technical Consultants* by Karl W. Palachuk, copyright © 2007.

Parts of this book are derived from or previously appeared in the *Network Migration Workbook, 2nd ed.* by Karl W. Palachuk and Manuel L. Palachuk, copyright © 2012.

ISBN 978-0-9763760-8-8

Library of Congress Control Number 2014941636

www.greatlittlebook.com

Electronic Contents

This book includes a few additional downloads which you will find very helpful. These include Word files, and a few other goodies.

If you purchased this book from SMB Books or Great Little Book, you should have received a download link when your purchase was completed.

If you lost that or purchased from Amazon or another reseller, you can register at www.SMBBooks.com.

Please have your purchase receipt ready to register. You'll need the Order ID. If your purchase somewhere other than SMBBooks.com, you'll need to forward proof of purchase to us.

Please respect our copyright and do not make unauthorized copies of these documents.

We welcome your feedback. Please email *karlp@greatlittlebook.com* or *danajg007@gmail.com*.

Warning about Used Books

When you register your book online, you agree that the book is no longer returnable for a refund. We simply have to assume that anyone who registers the book is going to download the electronic content and use it. Therefore, the book cannot be returned once the e-version has been downloaded. That also means that the owner of a used copy of the book does not have access to the electronic content. Thank you for your support and understanding.

Project Management in Small Business

How to Deliver Successful, Profitable Projects on Time with Your Small Business Clients

Dana J Goulston, PMP
Karl W. Palachuk

Table of Contents

About The Authors

Dana J Goulston, PMP was a project core-team member of the *PMBOK® Guide*, 3rd edition, and was co-lead of the Chapters 1 and 2 team. He wrote the majority of the sections related to the subject of PMO, as well as the glossary entries. As of this writing, Dana is also a member of PMI's Consensus Committee, which approves the release of PMI standards, new or revised. Dana has been doing Project Management since the 80's and has been a PMP since 2002.

Karl W. Palachuk has been an IT Consultant since 1995 and is one of the pioneers of the managed services business model. One of his books - *Managed Services in a Month* - has been the number one book on managed services on Amazon.com for more than five years.

Karl is a popular blogger among managed service providers and produces a wide variety of educational events each year, ranging from online classes, in-personal seminars, and the only all-online three-day conference in the SMB channel.

Acknowledgements

From Dana: Throughout my life, there have been many who have encouraged, pushed, prodded, and booted me along the passionate path called writing ... the path on which I belong. It started with my deceased paternal grandparents who promised me a dime should I find any grammatical or spelling errors in their summer camp post cards (I didn't make a cent), through the countless High School and College English teachers and professors dishing out their FULL range of grades, to today when my dear friend and co-author Karl came up to me at Karaoke one night and said, "Hey, let's write a book together."

From Karl: It's true that no project – especially a book – is successful because of just one person. This is my eleventh book and I have a very large collection of people to thank. In addition to Dana, of course, I thank the Band of Extraordinary Gentleman who were part of the book launch event when I released my first project management book: Arlin Sorensen, Matt Makowicz, Erick Simpson, Dave Sobel, and Stuart Crawford. Thanks to a major unified effort, we sold 120 copies of that book on the day it launched.

Those gentlemen have remained constant friends and sometimes business partners for the last eight years. And of course I knew them all before that. I am grateful to be associated with every one of them.

Section 1 - Overview

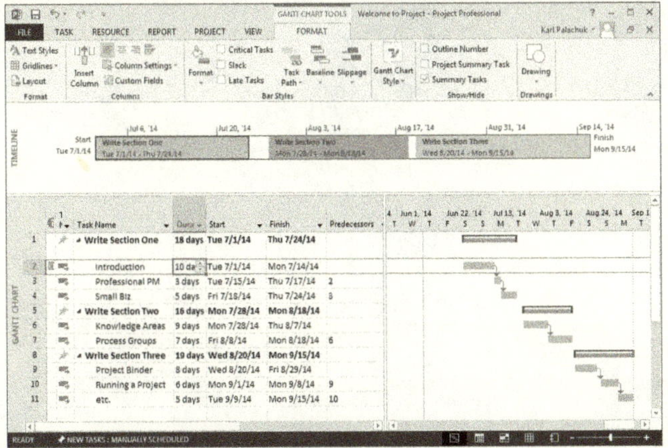

Chapter One: Introduction to the Book

Thank you for investing in this book. We trust you will find significant value here. Both authors have written extensively on project planning and project management. (See the "About the Authors" section.) But this Book is different from our earlier works in several ways.

We come from the two ends of the PM (project management) spectrum. Dana is one of the most prominent authors in the world of professional project management at the enterprise level. He manages projects for large corporations, nation-wide organizations, and generally "big" clients.

Karl has spent twenty years managing projects in the small business space. Where Dana dealt with massive projects with hundreds of personnel that lasted months or years, Karl's experience has been with projects that have a handful of personnel and last days or weeks. On rare occasions, months.

When we first met, project management was the most obvious topic we had in common. And almost immediately we conceived the idea for this book. Now less than two years later, here it is.

We see a need for more "professional" management at the small end of the PM spectrum. And, as it turns out, there are just a few things that need to be done to guarantee the success of smaller projects.

Note: Absolutely nothing derogatory or negative is implied by the use of the term “small.” To small and medium business (SMB) clients, a project is just as important as it would be for an enterprise client. The absolute dollar amounts are lower, but the relative costs are the same. In addition, business disruptions are just as important, as are positive outcomes.

So this book is focused entirely on the SMB (small and medium business) market. It is also focused entirely on technical or “information technology” projects. While it might be very useful for other projects in the SMB space, all of our examples are from technical projects. Finally, this guide is extremely practical. As with all of the books in the Great Little Book catalog, our focus will be on a down-and-dirty, get-it-done approach. If you want the history and philosophy of project management, Dana has other books for you.

Our primary audience is the technical consultant. Our secondary audience is the in-house information technology manager. If you are in another role and find this material helpful, please drop us a line and let us know that.

The Approach of This Book

For purposes of this book, you can consider a project to be any undertaking that requires more than **two steps** that can’t be completed at the same time.

So, for example, changing your password is not a project. But installing a new server is a project (discovery of old network, building of new server, install virus scanner, install backup software, etc.).

There are also small projects and large projects. You will occasionally have "macro" projects, which are really big containers with smaller projects inside. Please note that you don't need a project binder for every project. But as the complexity of a project increases, the need to keep track of it also increases.

In this book we provide a project planning and management process that is easy to learn and easy to teach to your assistants, fellow technicians, and sub-contractors. We address all the key weaknesses of project management and provide a process so that no work goes un-done, the project is completed within budget, and the project timing is tamed.

That means you have checklists. When everything's checked off, everything's done. It also means there's a built-in process for spin-off work that comes up as a result of the project. When some piece can't be completed as planned, it goes into the "additional work" section.

All of which leads to . . .

Each project has a **scope of work**. Everything inside the scope of work is part of this project. Add-ons and spin-offs are outside the scope, and therefore outside the project. They get written down, additional service tickets are created, and nothing gets dropped or forgotten.

As a result, your projects are successful and profitable. We know that because, by design, there's a scope of work and an allocation of resources (primarily labor). When you complete the scope of work, you complete the project.

You also avoid "scope creep" – adding chores to this project. Very often scope creep is simply additional work for no additional money. With this process, all that additional work becomes service tickets that are billable on another day!

The Project Binder

One of the key elements to success with projects, as we provide them here, is the Project Binder. The binder is important to your success because it's the ultimate random-access guide to your project. You can flip it open and find out where you are in the project, the next action step, how many stages there are, and all the documentation you'll need at the end.

We'll get back to the binder discussion in much more detail. For now, here's the view from 30,000 feet.

The Project Binder consists of a series of forms. They are designed to elicit a lot of the details you'll need as you work through the project. The great thing about a standard set of forms is that you don't forget to ask any of the key questions, and you don't skip any steps.

The Project Binder started out as a "generic project" we could use for a variety of jobs. After all, project management consists, in large part, of approaching each unique job in a consistent manner that could, theoretically, be reproduced for the next client.

One- and two-person companies don't have much problem with projects as long as one or both parties keep themselves involved and informed. But, when a consulting company begins to grow, they face an important challenge that must be addressed: How do you keep a project moving in the right direction, and profitable, when you have to coordinate it between several people?

The answer is: You need a tool that provides a way for everyone to come up to speed – quickly – on your project. Once you have that ability, you can have several people work on a project and keep it all moving in the right direction.

Now we're going to take a look at project management from two perspectives. Chapter Two introduces "professional" project management that leads to certifications and, potentially, the road to a career in PM.

Chapter Three addresses project management in the SMB (small and medium business) space. SMB project management is not "different" or wrong in light of professional project management. As we'll see, it is simply less complicated. But you still need to make money, stay on schedule, and attain the goals of the project!

Key to Your Success

This is the 21st Century: We assume that you are using a modern PSA or Professional Services Administration tool. It might be ConnectWise, Autotask, TigerPaw, or something else.

No matter how you do it, we'll assume that you use Service Requests or Service Tickets to keep track of jobs. You could use an Excel spreadsheet, if it works for you. But you need to use something.

No matter what you use, we assume you use it religiously. That means that everything gets entered into the PSA. Everything gets written down. You want to make more money in your business? That's easy: Keep better track of all the work you do.

Anyway, throughout this document you'll see references to *Service Requests* and *Service Tickets.* These are interchangeable. But they're not optional. ☺

See www.ConnectWise.com, www.Autotask.com, or www.TigerPawSoftware.com.

Your To-Do List for The Chapter

______ Browse through the rest of this book.

______ Find a PSA (professional services automation tool) or two. Pick one and begin working with it.

______ Find at least one binder so you can start building your binder when the time comes.

______ Download the content that accompanies this book. See the instructions at the front of the book.

Notes:

Chapter Two: Professional Project Management Perspective

What is Project Management?

Before we get too far, let's define our terms.

> ***"A Project is any endeavor that has a definite beginning and a definite end."*** *

The first question is usually, "What's Project Management?" The second is most often, "And why do I need Project Management on my project?"

And here are the answers: The Project Management Institute defines a Project as

> ***"A temporary endeavor undertaken to create a unique product, service, or result."*** *

And Project Management as

> ***"The application of knowledge, skills, tools and techniques, to project activities to meet project objectives."*** *

* PMBOK® Guide, 5th ed. page 3.

In other words, a Project is something that has a start and a finish (it's temporary). And for the purposes of this book, it's something in Information Technology, and related to small businesses.

We need Project Management for a number of reasons, and reasons that most of us haven't pondered. By employing a minimal amount of "management" or oversight to our project, we increase the probability of success by a HUGE amount. The payoff for the small investment is big-time. If we went to Las Vegas and gambled with theses odds, we would all be rich ... well mostly.

The bang for the Project Management buck is large. Basic Project Management is easy and doesn't take a large amount of resources to undertake. We want to be successful don't we? And most of us fall into the pitfall of not even thinking about the possibility that something might go awry on our project. We charge ahead, full of enthusiasm and excitement. We want that new server up and running. We want to play around and configure it so we can learn and experience something new and thrilling.

But caution is the order of the day. Yes, it's fine to mount the horse and ride off to do battle. First we must don our armor, sharpen our blades, strategize about our foe, and even get into shape. We plan and organize and strategize and think about the things that can go wrong as we fight the good fight. In this way, we prepare ourselves to the best of our abilities.

So we do a small amount of Risk Management, Stakeholder Management, Cost and Schedule Management, etc.

An added benefit is that we can now proudly show our boss or client that we know what we are doing. We can present documentation showing we did some due diligence in our preparation to spend their money. We can routinely report back to them on progress (they love this), and if something is going wrong, we can catch it quickly … and even before it happens, go to the one who pays the bills and present the situation … often with well thought out options and remediation plans (that's Risk Management friends).

By doing real Project Management on our small projects, we can:

1. Keep an eye on expenditures and the budget
2. Make some plans about the things that can or might go wrong on our project
3. Keep our employer or client in the loop regarding the progress of our project
4. Have a good sense as to the project schedule … and know where we are on the timeline
5. Know ahead of time what to expect at the end of the project (deliverables)
6. Be able to decide if we met our deliverables, and if not, why
7. Know what the others on our project are supposed to be doing, and be able to follow their progress … and be able to help out if needed. Have a head's up about anyone on our team who is planning a vacation, etc.

… And a bunch of other cool stuff!

The steps outlined in this book are taken directly from the Project Management "bible" – *The PMBOK© (The Project Management*

Body of Knowledge). And the bonus is that one of your authors – Dana – was an author of *The PMBOK©* (3rd edition). So You are getting this from the horse's mouth.

However, the above text is a comprehensive guide to performing Project Management activities. Neither of your authors, with all their decades of IT experience, and all their certifications have EVER participated or managed a project that included EVERY facet of Project Management outlined in the PMBOK. Every project uses one or some of the facets, but the book is designed to encompass all areas of Project Management, the bulk of which is not needed for our small business projects.

In that light, the sections to follow are trimmed down parts of PMBOK concepts. They are the barest pieces your authors felt would get you the best return on your time invested by reading on … and incorporating the practices. The sections however do follow the PMBOK organization of Project Management into what are called "Knowledge Areas."

In this book, companies are broken down and categorized by the number of employees, and in that light, the inclusion of the various knowledge areas follows. As an example, we felt that the technical expert undertaking a project, in a small business with less than 10 employees, really didn't need to concern herself with what's called "Human Resource Management." Larger companies do have to be aware of their team members' status, but not here.

However, we felt that EVERY project size should do some minimal Risk Management. The reasons will follow. Take a look at the chart discussed in the Knowledge Areas chapter, look at the size of your company, and see which Project Management modules are necessary for you and your project. Of course, these are guidelines, and if you feel a Knowledge Area is or isn't needed, feel free to adjust. The final deliverable is the target. Whether it be a new mail server, or an update to Windows Server. And know that these sections are put here to help you, to help your project, and to make you a success.

The Triple Constraint

There is a phenomenon in Project Management called the Triple Constraint. Picture a flat triangle of some material hanging by a string from its exact center. It's in balance, the triangle is held perfectly horizontal in space because all three corners are at equal distances from the center string and all three sides of the triangle weigh the same.

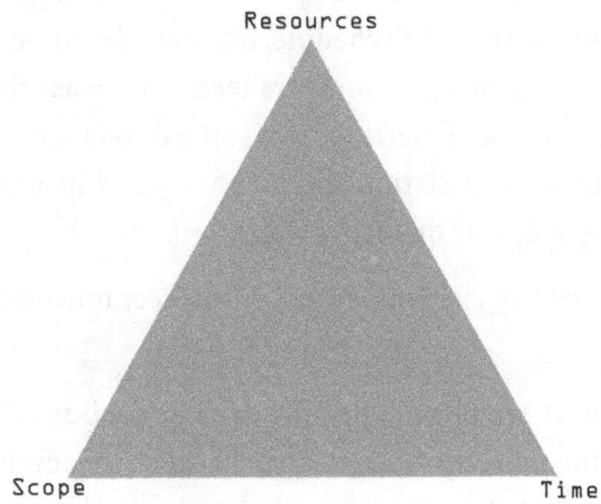

Each of the three corners of our hanging triangle can represent Scope, Time, and Resources. Some schools of thought call them Quality, Time, and Resources, and some even have a 4-sided representation. We like three and we like simple.

If we think of our perfect project with all three corners in balance throughout the project life cycle, the world will be good. Our project will be good. Our project sponsor or owner will be happy.

However, if we adjust or upset any one of the three corners, the other two will be affected in an opposite way. For example, if we cut our budget for any reason (lift the "resources" corner), the other two corners, quality and time will change. So if we cut our budget, we must either increase the time we need or reduce the quality we expect (de-scope).

Similarly, if our boss comes to us and tells us we need to have the server done 2 weeks before we had planned, we will need to tell Ms. Boss, "I'm happy to trim the schedule, but in order to do that, we'll need more money to bring on an extra tech." (increase the resources corner), or " we'll need to trim some of our server features, or fast track our OS installation and maybe even skip a testing cycle" (reduce the scope or quality corner).

See how the three key elements of any IT project must stay in balance?

Our job as Project Manager ... the one driving the bus ... is to bring the Project in on-time, on-budget, and (because we employ the philosophy of the Triple Constraint) with the expected results. By adopting the methods and descriptions of this book, our chances of achieving success in these areas will increase greatly.

Key Resource

If you are looking for a great "starter kit" on organizing your work and life so that you're always focused on the next action to be taken, read *Getting Things Done* by David Allen. You may not use his "system" exactly, but you'll learn a lot, and I'm sure you'll adopt at least some of his excellent advice.

"Professional" Project Management

If you poke around in project management for very long, you're likely to find references to the Project Management Institute and their certification program – Project Management Professional (See www.PMI.org).

Don't plan to rush out and get this certification tomorrow afternoon. The PMP certificate is serious, major, high-level stuff. If you have a bachelor's degree, you'll also need 35 hours of project management education and 4500 hours of project management experience. That's more than two years of fulltime work.

Most books and courses on project management are focused on the PMI (Project Management Institute) model and are intended to help you understand the concepts so that you can pass the PMI exam. **This book takes a different approach.**

If you're already familiar with the PMI model, nothing in this book will be too odd or out of place. But this book is intended for the SMB (small and medium business) consultant and our goal is much more "cut to the chase." Our single goal is to give practical advice so you can start having successful projects right now.

While you might catch the project management bug and decide that PMI certification is the life for you, it is much more likely that you enjoy being a small business consultant. You just want projects to go more smoothly and make more money.

There is a "lighter" exam by CompTIA called Project+. See www.comptia.org for more information. While not as rigorous as the PMP exam, the Project+ exam is based on the same methodology and terminology.

A Note on Certification

Many professions or trades have certifications. These are validations from governing groups, regulatory bodies, or professional societies verifying some level of proficiency in the given field. They usually require some combination of professional experience, education, and having passed an examination. Common examples are CPA (Certified Public Accountant), Esq. or Esquire (those lawyers licensed to practice law), and AIA (Member of the American Institute of Architects). Certifications are not to be confused with actual licenses themselves ... which are usually issued by governments allowing the practice in that jurisdiction.

In US Project Management, the Project Management Institute (PMI) in Pennsylvania is the body that issues and tracks professional Project Management certifications (there are several levels). There are several international Project Management bodies, but for the most part the PMP is an internationally recognized certification.

The PMP (Project Management Professional) is the commonly sought-after certification by those looking to hire a Project Manager. While not a requirement by any means, it does show that a person has some education, experience, and knowledge of the field. In the past, all a person had to do was pass the rigorous (4-hour) exam and they would be granted a PMP certification. That worked nicely for those who were good at taking tests and was a challenge for those of us who didn't "test well." What that system did was dump a bunch of inexperienced people onto the PM market causing a degradation in the validity of the certification.

PMI has changed their ways and now requires a matrix of education/continuing education, documented real world experience in the field, and the exam. Today, more and more organizations are realizing that having the certification is a value. And as PMI continues to fine tune and self-evaluate the certification process, you can start to get a sense that the certification has greater and greater value.

It is not uncommon that state governments and large corporations require that large projects (say over $1 Million) be run by a certified PM, and that even larger projects have at least two PMPs.

All that being said, having a PMP certification is by no means a guarantee of Project Management expertise or proficiency. And as far as small businesses are concerned, even less important. However, if a Project Manager does possess this "stamp of approval," you can at least rest assured that they have some idea of what's going on in the field, and that they have managed projects in the past.

A major part (but not all) of the exam is based on a book called, "The PMBOK Guide" (Project Management Body of Knowledge). This tome is carefully reviewed and updated every five years as mandated by the universal standards body, ANSI (American National Standards Institute), an organization that oversees many technical and professional standards. One of our authors (Dana Goulston) was a co-author of the PMBOK Guide, 3rd edition (2004), and as of this writing, is currently a member of PMI's Consensus Committee, a voting group that approves any new PMI standard.

Your To-Do List for The Chapter

______ Buy and read *Getting Things Done* by David Allen

______ Check out the Project Management Institute and their certification program –See www.PMI.org

______ Check out the CompTIA Project+ certification. See www.comptia.org

Notes:

Chapter Three: Project Management for SMB Consultants

Controlling Information

It is very important that you have a process for controlling information. It doesn't have to be the process outlined here, but it needs to be *some* process. What do we mean by controlling information? There are several elements.

- How will you track the project?
- How will you track information within the project?
- How will you communicate these arrangements to clients and staff?

For example, here's how the Project Binder might work for you. You should be able to open the binder and instantly know where you are in the project, the last action taken, and the next action needed.

Tools

There are two key tools you should use in any project, and one tool you probably don't need in the small business space. The tool you probably don't need is project management software. The tools you do need are a Professional Services Automation (PSA) package and a Network Documentation Binder (NDB).

You'll find a variety of project management tools, including the big daddy of them all, Microsoft Project. Perhaps the great strength of Microsoft Project is its scheduling ability. Again, this is high-end stuff and you may find it useful for certain projects.

But most SMB projects aren't overly complicated, don't take months, and don't involve coordination of multiple teams.

Many Professional Services Automation (PSA) packages have project modules. Examples include ConnectWise and Autotask. While you may find these useful for tracking the action steps of your project, you will still need a written plan and a way to manage the look and feel – and profitability – of your project.

The bottom line for SMB is that projects just aren't that complicated. You need to take them seriously and do them well. But we're not building any aircraft carriers here!

The other tool you'll need is a good Network Documentation Binder – NDB. We'll call it an NDB even if your documentation exists only electronically. In addition to the Project Binder we'll discuss in the next few chapters, you need a place to keep thorough documentation of everything you do. When you're finished with the project, this documentation will be a permanent part of the client's overall documentation.

For Karl's book on network documentation, see www.NetworkDocumentationWorkbook.com.

Key to Your Success

Certification versus Practical Knowledge

We've heard so many arguments about this, we can recite both sides in our sleep. Here's the deal: Look at the reality of what you do. Ninety-nine percent of the consultants in the SMB space are never going to organize a project that has dozens of teams and an administration budget to handle the management of dozens of meetings.

In the real world, most of us will spend most of our time on very similar projects.

We're going to migrate networks to a new server. We're going to help a client change ISPs. We're going to install a new firewall in an existing environment. We're going to upgrade an office suite for all the desktops in a business.

So, for the most part, we don't need to learn the terminology of the Project Management Professional exam. And if you took all that super-structure and loaded it on top of a new router installation, you'd turn a $500 project into a $15,000 project. That's not good ROI (return on investment).

So the bottom line is: you should learn what you can about professional project management, but temper it with a healthy dose of reality. Most SMB consultants need to do more project management than they currently do. At the same time, most don't need a degree in project management.

So Where Do You Get Started?

As a business manager (or perhaps owner), you are probably concerned with maintaining quality as your organization grows. The problem is, how do you make sure that a job is done "the right way" every time if you don't personally do it?

Making your business grow means overcoming fear. Fears about money, clients, losing control, etc. One of the biggest fears is that, if you don't do the job, it won't be done right.

So go figure out what you're going to do about fear. But have faith in this important truth: Every company with more than one employee has overcome fear and figured out how to make things work. The bigger they are, the more fears they've overcome.

If you haven't read *The E-Myth Revisited* by Michael Gerber, go do that this weekend. Gerber points out that you need to create standard operating procedures. Write down everything that needs to be done and document how to do it. You can see why we like this guy!

But he doesn't lay out the specifics of how you create a procedure or process that works. So we're going to work on that in this book.

Closing the Loop

We're going to work with processes that are *closed loop*. That means that they are designed so that every task is completed or determined not to be necessary. More importantly, new problems that arise are either dealt with, scheduled for later, or determined not to be necessary. But nothing gets missed. Nothing gets dropped. Nothing goes undone.

Things outside the process stay outside the process (e.g., we only work on one service request at a time).

Here's an example. Let's say you have a New PC Checklist for your client, ABC, Inc.

Checklist:

- o ABC uses public folders for company contacts, so they need that set up on every desktop
- o Many desks are shared between users, so they each need a profile ready to go when they sit down
- o Small groups share specific printers, but everyone needs access to other printers
- o Everyone needs the company LOB (line of business software) shortcut
- o Finance and sales people have specific security groups
- o There are eight sets of email distribution lists, depending on job classification

- New PCs always go to managers and manager PCs are handed down to new employees
- Everyone gets Word, Excel, and Outlook. Some groups get Access and PowerPoint
- Sales people have a different format for their email address
- Everyone needs Adobe reader, anti-virus, and anti-malware
- List any items not done and why
- If you come across other issues, create a separate service request

The point is: You need a checklist! Can the customer set up their own computers? Yes. Will they do everything that needs to be done, in the right order? No. Absolutely not. Why? They don't have a checklist and you do.

You can guarantee that every workstation install or migration is handled the same way every time. Period. How can you guarantee that? Because your checklist is a closed loop process.

Notice those last two items?

- ✓ List any items not done and why.
- ✓ If you come across other issues, create a separate service request.

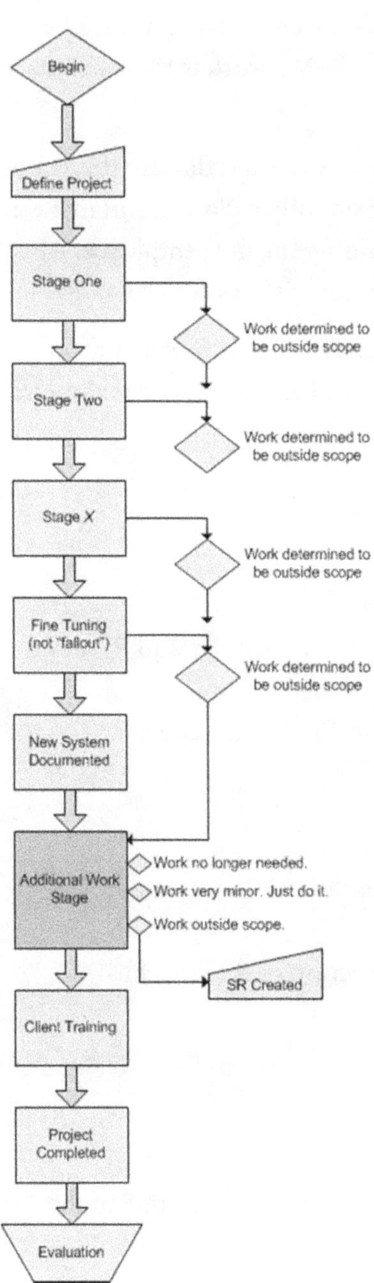

They're key to your success. If anything "interesting" comes up, technicians tend to just jump in and fix it. But if that something turns out to be a major project, then the original task isn't completed in a timely matter, and you have a tough time separating the two jobs for billing. If you've quoted a flat fee for the job, then you've really messed things up.

The *process* of completing everything inside the scope and separating everything outside the scope will set you free. It will make you more profitable. It will make clients happier because they see that Job A got done as promised, even though we now have Job B to tackle. This is much better than starting Job A and having it morph into Job B, and now neither one is very done at all.

Big whiner complaint:

"Every job is different."

We hear this all the time. "I didn't create a checklist or a form because every case is different." Yeah, yeah. My work is more art than science.

Ninety percent of all new PC setups in America today are the same. Another 9% is the same within any client office. So you *can* make a checklist from opening the box to training the new employee. And that list will cover 99% of what you're going to come across.

But, you argue, stuff always comes up. True enough. And part of the *process* is to write that stuff down, and either do it or make sure it is dealt with.

The broad outline looks like this:

1. Stage One tasks

- Do this section of the checklist
- Update the Network Documentation Binder and PSA system as needed
- Note any issues in the Issues section of this checklist

2. Stage Two tasks

- Do this section of the checklist
- Update the Network Documentation Binder and PSA system as needed
- Note any issues in the Issues section of this checklist

continues until . . .

x. Stage x = Issues Section

- For each item you have entered here, perform one of the following

- Fix it (minor issues). Make a note on the checklist.
- Create a new Service Request
- Mark issue as "no longer an issue" and note reason

Until you understand why this is a closed loop process, don't argue with it or say "but . . ." – Just do it.

In this process:

- ✓ Every task will be completed. If not, there will be a note in the Issues section
- ✓ No new tasks will be introduced to the process. New issues will be noted in the Issues section
- ✓ If you (and your techs) never check a box until a task is completed, then when you're finished, every piece of this process will be completed
- ✓ All issues in the Issues section are either completed, moved to a separate service request, or no longer needed
- ✓ Nothing goes undone
- ✓ Nothing is forgotten
- ✓ Nothing becomes a "stopping point" that keeps you from completing the process
- ✓ Nothing can break this process

If you go outside the process, then you worked outside your standard operating procedure. The process is still not broken. It is just not properly followed. ☺

Developing Checklists and Reusing Projects

Here's a sample checklist so you can see how the process works.

<u>Procedure: Deleting / Retiring Users From A Network</u>

Background:

People leave. Some clients have very stable personnel. Others have lots of turnover. In most cases, when people leave, they leave email accounts behind. The following procedure is designed to create a smooth transition without losing any email.

This process may be customized per client. Here's the checklist:

1. Disable the account. Note: Exchange mailbox backups in some backup programs, including Veritas Backup Exec may "fail" due to an inaccessible directory. This will clear itself in time as normal Exchange maintenance is done, or you can remove that mailbox from the backup.

2. Re-point any SMTP addresses according to the client's needs. For example, you might point the old user's email address to the department manager's mailbox.

3. Try to wait at least one day before continuing with any other steps to allow the client a little time to consider how they wish to proceed.

4. User Mailbox and all sub-folders get exported to

\main-public-data-share\ArchivedEmail\%username%

4.1. Be certain all folders are included in the export.

4.2. Be certain the .pst is not set to be password protected.

4.3. Use the outgoing employee's name for the .pst file. Do not use "Outlook.pst" or a similar name.

4.3. Assist client in consolidating any of the user's data.

5. Prepare for the data move by creating a directory with the user's name in the

\main-public-data-share\archivedUsers directory

e.g. \\ASSBS\Public\ArchivedUsers\JoanS

5.1. Move the data (do not copy) onto the server from:

- The user's workstation Desktop
- The user's workstation My Documents folder
- The user's home folder on the server. Note: Do not move My Music, My Pictures or My Videos (etc.) unless explicitly requested by the client.

5.2. Set permissions on the data according to the client's needs.

6. Clearly instruct the client that as users find data they need in the archived user's data directory, they should move it to

the best location for that data, such as the proper location in the public folder, the associated team folder etc.

7. If a user needs access to the old .pst file for the departed employee, make a copy of the .pst and then have the user open the copy. That way, any changes, deletions, etc. happen on the copy. The original is retained as archived data.

8. Remove the user from the domain and be certain the checkbox is checked to "Remove user's home folder."

8.1. Run the Exchange mailbox cleanup agent and then purge the user's mailbox from Exchange.

Your process may vary, of course, but the key to success is to *have a process.* Your process may be exactly as described here. But if it's not written down, even you will not execute all of these steps in order.

And when you hire technicians who are NOT the owner and don't have your attention to detail, then there's no way they can follow this process. It has to be written down and they have to be trained.

Concluding Thoughts on Project Management for SMB

The discipline of "project management" didn't come into existence because academics decided to tell people how to get things done. It was developed by people who wanted to manage complicated tasks and make sure everything was done, done right, and completed within a reasonable budget.

If you look at the history of old cathedrals and castles, you'll get a sense of what the world was like before project management. There was (usually) an architect. But no one knew how much the project would cost. It got started, but stopped when the money ran out.

When it was time to pick up the project again, no one knew what had been done and what remained to be done. Even if they had plans, these were scrapped in favor of a new idea. Sometimes towers were added because the king (or whoever was paying for it) wanted a tower.

The result is a building with one square tower and one round. Turrets on one side and not the other. Over "budget" because there was no budget. Sometimes these buildings sat half-built for centuries before someone decided to put up the money to complete the job. And it looks like it!

If you've ever been involved in software development or database programming, the analogy to building a cathedral is very clear.

Here's how a bad project goes:

- When you start a project, everyone wants "the best"
- You suggest a "discovery" stage to define exactly what the outcomes will be and how much it will cost
    - This is rejected because it costs too much
- The "project" begins by simply starting to do work
- Then they discover that the best is really expensive
- Some key features are cut. Shortcuts are taken
- Frustration mounts as the client realizes that key features are missing

- Some work has to be redone several times because it was done in the wrong order, or was done based on a plan that changed. Or perhaps a feature that was removed is now re-added.
- Eventually, the "project" devolves into a series of requests to continue patching the "system" you've built. There is no ending. There is no conclusion. There is no evaluation of the process. So there is no learning for next time.

(If you've never had this experience, then we assume you've been in the consulting business less than a month. Welcome to the profession! You'll need a Phillips screwdriver and a label maker.)

But don't fret: Life got better.

Over time it became clear that properly formulated projects could be repeated. Two buildings could be built the same. Two guns could be built the same. Two ships could be built the same. And ten ships, and ten thousand cars.

So, you see, project management grew up for very practical reasons. And, for the most part, they are the same reasons you need to manage projects:

- You can predict how long a project will take
- You can predict how much a project will cost

- You define goals, vision, and a purpose (and therefore you can measure your success)
- You can define what the end result will look like before you start
- There are fewer surprises and fewer disappointments
- And the entire process can be repeated again. In fact, if you've done your work properly, it will be better the second time!

You'll also find other benefits that make your business more profitable:

- You'll have better communications with the client
- A project can be handed off from one technician to another
- If a key person leaves (vacations, illness, moved to a new job, got hit by a bus), the project can continue and be successful
- You can reduce downtime significantly
- You can use your project management approach to make sales! Does your competition have a system for managing projects, or do they just show up and start working?
- You'll make more money as you repeat similar projects with different clients. After all, you'll fine-tune the process each time. So it just gets better and better.

There's a cost, of course, to managing a project rather than just "doing it." You'll have to sit down and go through the binders. You'll have to keep track of where you are with each project. But the truth is that this is something you should be doing already. The binder process will just make it easier.

And you'll always know – or at least have at your fingertips – where you are and what's next.

Your To-Do List for The Chapter

______ Buy and read *The E-Myth Revisited* by Michael Gerber

______ Define your procedure for removing an end user client from the network. Fine-tune it for your business.

______ Buy and use a book on Network Documentation. One option is at NetworkDocumentationWorkbook.com

Notes:

Section 2 - Project Methodology in the Small Business Space

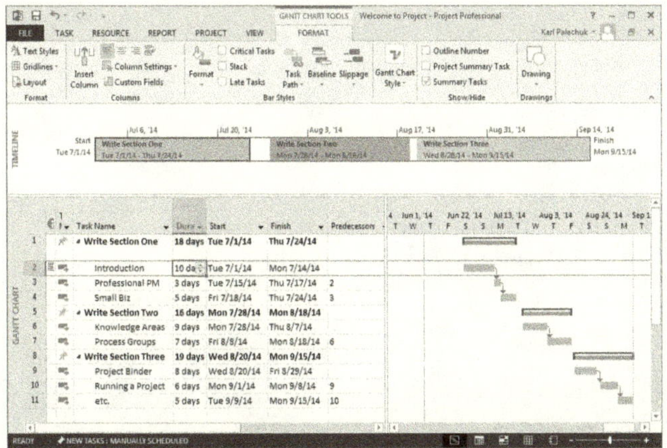

Chapter Four: Project Management Knowledge Areas

The PMBOK Guide breaks the management of projects up into nine easy-to-digest pieces called "Knowledge Areas" or "KA's." In a perfect world with infinite resources, each and every one of the nine KA's would be used. For this text, and our small businesses however, some of them are overkill, some of them are needed in part, and some of them really should be used in their entirety. In this text, we've identified three flavors of project teams with each flavor utilizing differing combinations of the nine KA's.

Notice the chart below: Group C is for organizations with more than 50 users, computers, or "seats." Group C is mostly out of the scope of this book, but if you're reading this book, and your company falls into this category, follow the recommendations for Group B keeping an eye on "Human Resources Management" in case your company has formal policies and procedures regarding the acquisition, formation, and usage of your project team. This is especially true if your organization has reached any level of "Project Maturity," has a formal Program Management Office (PMO), or has a "matrix" organization.

- Group A is all project teams that support organizations of between one and ten employees

- Group B is all project teams that support organizations of between 11 and 100 employees
- Group C organizations, as mentioned above, have 51-100 employees

As an organization grows, the need to introduce more formalized Project Management practices grows with it (if one wants to be successful), and as such, Group B organizations really should perform a few more things as they manage their project endeavors than Group A. The following sections outline which of the nine KA's are necessary for both Groups.

Component required			
Group	A	B	C
Employees/seats	1-10	11-50	51-100
PM overview	Yes	Yes	n/a
Integration	No	No	n/a
Scope Management	Yes	Yes	n/a
Time Management	Yes	Yes	n/a
Cost Management	Yes	Yes	n/a
Quality Management	Yes	Yes	n/a
Human Resources Management	No	(as needed)	n/a
Communications Management	No	Yes	n/a
Risk Management	Yes	Yes	n/a
Stakeholder Management	Yes	Yes	n/a
Templates	Yes	Yes	n/a

Each organizational type should perform the following steps (the degree to which and the descriptions will follow):

Summary of Project Management Knowledge Areas by Group

Group A: (1-10 employees)

Scope Management

Time Management

Cost Management

Quality Management

Risk Management

Stakeholder Management

Group B: (11-50 employees)

Scope Management

Time Management

Cost Management

Quality Management

Communications Management

Risk Management

Stakeholder Management

Now what do these things mean? We shall tell you, but please don't run away. Each section will contain a simple checklist telling how to use them. For now, find your group by the number of employees you have. If you're on the border, choose the lower group type. Then just check off each component to be worked.

(Note: We've made note of where each Knowledge Area corresponds to chapters in the *PMBOK Guide*, 5th Ed.)

Scope Management (see PMBOK, Ch. 5)

Scope Management is all about deciding what the scope of your project is, or exactly which "things" your project will produce or deliver (these are called "Deliverables"). These are the exact items or features we (and our sponsors) want … what they expect for their money. Scope Management is about putting measurable parameters around those deliverables (Scope) and not letting things get away (Management). In other words, deciding EXACTLY what you want to do on your project and sticking to it.

When deciding on what the project is going to deliver, be specific. Suppose we're having a party and need some cheese from the store. And just suppose we know what we want ... maybe we're pairing a certain type of cheese with a certain type of wine (or a certain type of Hard Drive array with a certain type of server). We don't send our friend off and say, "Go get some cheese." We would say, "Go get 3.0 pounds of Point Reyes Blue Cheese (+/- .25 lb.)." We want and we need measurable, testable, deliverables on our project.

How many times have we done something and as things go along, we say, "I can easily add this piece of code or hardware accessory because I'm here and this would be a great time to do that." Or "so-and-so would love it if we added this or that."

We have one piece of advice for that: **Don't Do It!**

And here's why. Adding "good guy" stuff (also called "Gold Plating") introduces variables to the project. It's more things that might go wrong, it increases the budget (even though it doesn't seem like it at the time), or maybe even things the sponsor just doesn't want. It's a risk not worth taking despite how good it makes us feel at the time. Just Don't Do It!

When we clearly specify and document the discreet deliverables of our project, we can accurately assess whether or not we've achieved our goals. We can also avoid ambiguity and misinterpretation, cost over-runs, arguments and finger-pointing, and even legal action. We clearly define our project ... and *then* we get all interested parties to agree and sign off (yes, have a short signature form ready). Everyone is then on the same page, and all expectations are in line and achievable.

Checklist – See related forms for this later in the book

- ❑ Collect requirements (this might mean documenting what is needed on the project)
- ❑ Define the scope (list the discreet things your project will achieve)
- ❑ Break the work down into bite-sized pieces so you can measure and track your progress

- ❑ Document and get agreement on what the outcome of the project will be (deliverables). Get a signature on a piece of paper.
- ❑ Stay on top of it ... manage it (aka Project management)

Note: all things change, and the scope of our project is certainly no exception. No worries. If both parties agree on a change, figure out the cost, make an adjustment, document it, re-sign the changed document, and *manage it.* Later we'll talk about adding items to the project vs. create new work orders outside the project.

Time Management (see PMBOK, Ch. 6)

Time Management deals with deciding how long the actual pieces of your project (the work) will take and doing your darndest to stick to your time frame. Obviously nobody has a crystal ball ... and how many times has a client come to us and said, "How long will this take?" But if we've done something enough times, know people who have done things similarly, or even done it once before, we can make some rough estimates. Further, as we progress down the work path, we learn about our environment and where our work is going, our estimates will get closer and closer to the actual time things will take (called convergence).

We should take an educated stab at the time frame before we begin to write it down, and over the course of our project (preferably at regular intervals), rinse and repeat. We will then have a fairly reliable time frame.

These estimates can be done in a few minutes and really won't add much work to our plates. The payback will be worth it.

PM Bonus: if we multiply our estimates by the hourly rate of those doing the work, we might actually arrive at an initial cost estimate ... DING! Then we repeat this costing exercise at regular intervals and we get a very good idea of our "burn rate" or at a minimum whether or not we will be on target or over budget (nobody is ever under budget). See the next section on Time Management. The two are tied together.

Cost Management (see PMBOK, Ch. 7)

Few have millions of dollars, or infinite sources of funding. That's why our businesses are small. We're usually starting up with very tight budgets, and often, those paying the bills (or maybe that person is us), are keeping an eye on almost every penny. For that reason alone, monitoring our spending is both prudent and is also a solid PM practice. If we're doing hardware installations, we can get extremely accurate costs of the materials involved. And as noted above, we can also get some decent idea of how much our resources (people) will cost the project. A few fudge factors (unforeseen delays or setbacks) and some additional stuff like people getting sick or taking vacations, and we can establish some budget baselines.

A good rule of thumb is to add 10-15% for labor costs. It is our strong recommendation that you track all time – even your own. However, if you are the owner or a key player in a small business, we can justifiably not add your time to the budget. Just keep in mind that you won't be available to do other company stuff.

If we have some techs, or if we have to bring in some outside labor or consulting assistance, we might go on the high side.

If we can get written quotes and subsequent contracts for the hardware all the better.

The bottom line is to get an idea of our project cost in total and do our best to stick to it (manage it). It's in everyone's best interest.

As mentioned in the note above, it's our job as Project Managers to keep an eye on the budget throughout the project's life cycle. We must, at a very minimum, inform our sponsor or project owner if we suspect the budget will go awry as soon as we are aware. We must separate our emotions and any notion of personal responsibility for a budget overage (whether or not we are responsible) and present the information in a calm, professional manner. We should also have suggestions at the ready.

Budgets *do* go over. It's a fact of life regardless of how good we estimated the budget at the outset, or during the project. It happens. Prices change. People are human, make mistakes, and change their minds. They even quit or go on vacation, have babies, and do all that other stuff that affects our project's budget. Just be aware of things when they happen, don't pretend they didn't happen and hope they will go away, and deal with them in a professional manner.

[Dana's rule of professionalism: "Being able to separate our personal issues and do the job at hand."]

If we keep an eye on the budget, our burn rate, ongoing changes to our project, and stay in professional and constant communication with our key stakeholders, our project will be successful. Who could ask for more?

Quality Management (see PMBOK, Ch. 8)

Normally, in larger organizations, Quality Management is concerned with understanding any and all quality standards your organization might have as they relate to your project. For our purposes, and in most small businesses, there are few if any formal quality standards implemented in our organizations (ISO, ANSI, IEEE, EIA, etc.) However, we must first inquire. If there exists ANY quality standard in our organization, we MUST understand them and incorporate them into our project. What this also means is that we must document how our project's deliverables conform to those standards.

If there are no quality standards in our organization, it's still important to document any quality concerns or measure we choose if we implement them into our project. Sometimes the effort therein is minimal up front, and will have benefits down the road in the form of risk management, or even quality marketability as we go forward with future projects. Another added benefit is that if we implement any quality standard … either formally or informally, we have the added assurance that our project has that much greater probability of succeeding on the deliverables we've determined.

Note:

"Quality Assurance" is the measure of the quality of the **process** while

"Quality Control" is the measure of the quality of the **product**.

Communication Management (see PMBOK, Ch. 10)

Communication Management involves deciding and following processes on our project that relate to communicating what is happening during the entire project life cycle.

This area cannot be stressed enough. Proper and timely communication can make or break our projects. Unless we are doing the project by ourselves, and for ourselves, we MUST keep others in the loop somehow.

Starting at the top, we need to keep our sponsors informed about the progress of our project, the preliminary budget and subsequent expenditures, any and all challenges we might encounter (expect this ... it's foolish to think that ANY project will run perfectly from beginning to end). The key is proper planning and continued communication.

We also need to keep our project team informed as to what is going on and how things are moving forward, or what challenges are being faced.

- The developers need to know if their environment will be ready as planned
- The sponsor needs to know that their server delivery will be delayed
- The users need to know that their way of operating may change, or that their favorite application will be getting an upgrade, and when their systems might be down

- The suppliers need to know if we decide to add a feature to our order

You get the picture … Communicate Communicate Communicate!

All that being said, how do we "plan" for communication? This is easier than one would think. First make a list of everyone who needs to be in any loop. You might ask your sponsor or team members to jump in here in case you forgot a key player. Then think about what they need to know, and when. Ask the sponsor how often she wants to receive a status report. If it's a short duration, will it be daily … weekly … hourly. Document the list before the project starts and get "buyoff" from those on the list to make sure it's sufficient to meet everyone's needs.

Keep in mind, some participants might think they need more communication than you do. This happens often in micro-management environments. If you disagree, have a conversation with them and talk about the value both to you and to them of more frequent communication. If still unsure, check with the project sponsor. Sending an end-user daily status reports might not be the most appropriate communication route, and certainly not the best use of your time.

Dana was recently on a *huge* multi-site, multi-application deployment that lasted 5 days and cost the client several million dollars. The sponsor and client wanted a status report every FOUR hours! It was a challenge, and he did have to set his alarm and touch base due to 24-hour shifts, but it was required.

If you need to formalize or keep track of who gets what, there are many free communication templates available on-line. We highly recommend a PSA tool for tracking this stuff. But you may also want to simply rely on an Excel spreadsheet to track communications. A sampling of these is listed in *Appendix C: Resources.*

Risk Management (see PMBOK, Ch. 11)

Wow! Risk Management! What in the world is that? And do we really need it in our modest IT projects? The answer is an unequivocal YES! Let's talk about what it is and why we need to do Risk Management on our projects.

Risk Management is the institutionalized art of worrying.

Those of us who are Risk Managers have turned worrying into an art and a science. Dana has even taught others how to worry for a living. Actually ...

Risk Management is deciding *ahead of time* the things that might go wrong on our project so we can do one of two things:

a) Do something **NOW** (or soon), to reduce the chance that it might happen

or

b) Do something to minimize the harm if something does happen.

We call those things **Probability** and **Impact** respectively.

Now here's the art of it all. If we gather our team together (preferably before our projects starts) and make a list of all the (realistic) things that can go wrong, then we can rank those things in order of nastiness. After we have our "prioritized" list, we can take some actions NOW for say, the top 50% of the things on our list (you decide what you want to tackle ... what makes sense to you on your project.)

This systematic method reduces the chances that if and when our project encounters detrimental things, we've done something about it to minimize the damage or we've reduced the chance that the detrimental thing happens to us and our project.

This type of preventive thinking, planning, and acting is prudent and professional as well as diligent. By spending some time before the project we can make some educated assumptions and preventative measures to enhance the likelihood our project will achieve success. And we all want success ... right?

So many of us do minimal planning and want to charge ahead and "git 'er done." While this method (sometimes called the "Cowboy" method) is gratifying and satisfies the urge to see results, it isn't the best way to approach our projects.

Let's see exactly what it is that we can do in our Risk Management step:

1. Risk Identification
    a. This is the first step in figuring out which things can negatively affect our project. These are the unforeseen ... or maybe "we know but choose to ignore them" gotcha's. Know this: *no project is free of things that can go wrong*! No project is perfect from start to finish. And even if there is such a critter, and even though you have this fantastic book helping you through your project,

there are things that we at least need to consider for a minute. So what are they?

- Get your team in a room with a white board and some pretty markers
- Write down everything you can think of ... everything and anything that can possibly derail or cause harm to your project. Remember these are things that may or may not happen *in the future*!

  If something is in play, then it is not a risk: It's called an Issue, and it's dealt with after the fact, or retroactively. Good Risk Management is preventive – we do things before the bad stuff happens.

  Examples: The server vendor won't deliver as promised; the drive array will fail; the computers will draw too much current and blow a circuit breaker; the heat from the servers might build up and make the room too hot; the client's check will bounce; the client will change their mind or want to add some "features."
- Capture all of these with no criticism as they come forth from the neurons. Save the sorting and all that stuff for later. Right now we want to tap into our creative right-brain.
- Now go away and type all these things out on a simple document. Go get lunch.

- After lunch, assemble the team for part II ... the triage, or prioritization (also called Risk Analysis).

2. Risk Prioritization or Risk Analysis
    a. Go down the entire list and have discussions. The list will sort itself as you decide what's important, what's silly, what's in the middle. Maybe you can give them some ranking values. If you do, here's our suggestions:
    b. Give each a value of the **probability** of the thing happening ... High, Medium, and Low (H, M, L).
    c. Also give each the same value of the **impact** to our project should that thing happen (H, M, L).
    d. Now take the two values and put them next to the event (HH, HM, MH, MM, ML ... LL).
    e. Now sort! The HH items go on top. These are the risks which need the most attention. After are HM, MH, MM, LH, HL. (these risks need moderate attention). Last are ML, LM, LL. These are things which need little attention ... but we must at least pay attention to them during our project. See the tables below.

Probability:	High	Medium	Low
Impact: High	HH	MH	LH
Impact: Medium	HM	MM	LM
Impact: Low	HL	ML	LL

Sort Order:

HH

HM, MH, MM, LH, HL

ML, LM, LL

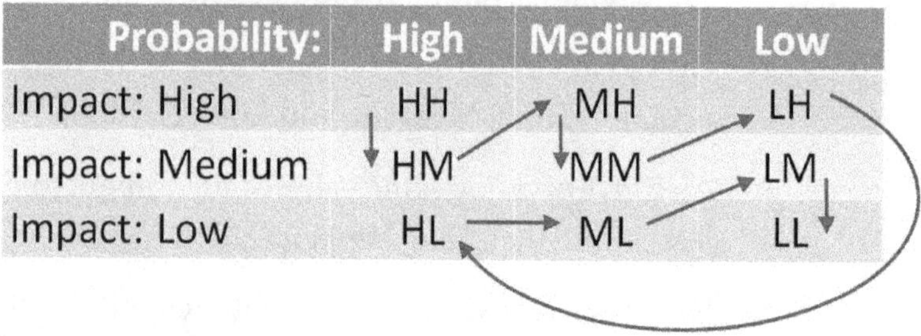

Sort Order

Probability:	High	Medium	Low
Impact: High	HH	MH	LH
Impact: Medium	HM	MM	LM
Impact: Low	HL	ML	LL

Note: feel free to use whatever scale you like: 1-5, 1-10, etc. As long as it's a scale and you can combine the two values in some way to arrive at a logical prioritization.

3. Risk Mitigation

   Next step: Look at the items on the top of your prioritized list. We need to do something to "mitigate" these risks (the PMP term is "remediate," the common term is "mitigate"). We need to do one of two things (two is better, but one at least):

a. Do something to **lower the probability** of the thing happening

b. Do something to **lower the impact** should the thing actually happen

For example: Let's say you are about to migrate email from an in-house Exchange server to hosted Exchange mailboxes. What are some problems you might run into?

- Huge mailboxes
- Downtime
- Corrupt mailboxes
- Failed backups
- We don't control the DNS
- Coordinating active directory

Which of these can you prevent (reduce the probability of)? Perhaps cleaning out mailboxes and getting them all to a reasonable size is the answer.

Which of these can you minimize the impact of? Perhaps work with the ISP to make sure that you can get DNS changes made, propagated, and tested in a timely manner.

Stakeholder Management (see PMBOK, Ch. 13)

What is a *stakeholder* you might ask? No, it's not the guy holding the Rib-Eye's over the Barbie. The PMBOK defines stakeholder as any individual, organization, or group who may affect, be affected by, or perceive itself to be affected by a decision, activity, or outcome of a project. What does this mean? In broad terms (for the pedantic among us), it can be anyone who has an interest in the outcome of our project (be it good or bad).

For our purposes, it's really the main players who have the biggest interest in the success of what we are trying to do. Our primary client contact, the company owner, the main users, even ourselves if we are key individuals in the organization (the single IT guru in a 50-seat small biz). It almost always is the person paying us, and it is almost always the users or primary users of our servers, software, PC's, or printers.

We could call everyone who has anything to do with our project a stakeholder, but for the purposes of getting the bang for the buck and picking the key individuals who we want to keep happy, we refer to them as key-stakeholders. It's up to us to decide who the key stakeholders are, but it isn't that hard to narrow it down to a handful of people who we *really* want to get their needs met.

So stakeholder management consists of:

a. Decide who the key stakeholders are
    - Make a spreadsheet or a list
b. Decide what each of these key people needs or expects out of our project
    - Write down what each individual or group expects from the outcome of our project

c. Decide how and when you will communicate with them (weekly emails, a knock on the office door on Fridays, etc.)
   - Write down how often you will communicate with each key stakeholder
   - Write down what you will tell them

d. Keep an eye on each and make sure they are happy (not an easy task when things go wonky ... or over budget)
   - Make a note about what to do if they seem unhappy or express concern

Keep It Simple

Don't get too carried away worrying about Knowledge Areas. Really, these are just the common sense things you know you need to take care of to make your projects successful. In fact, one way or another, you've already been managing these KAs – you just may not have used that terminology.

Later in this book you'll see that we have addressed the final KA in fine detail: **Templates**.

Creating checklists, templates, and "canned" projects can save you lots of time and money. In fact, when you can execute similar projects consistently time after time, it's fair to say that templates will make you money. The more you use them, the more consistently you'll be able to quote projects, win projects, and execute projects – on time, on budget, and successfully reaching all goals.

Your To-Do List for The Chapter

______ Make a list of 3-5 projects that you execute again and again (e.g., migrating email to the cloud, migrating from SBS 2011 to hosted services).

______ Since Communications Management is the primary Knowledge Area difference between Group A projects and Group B projects, begin drafting guidelines for when you will include Communications Management in your project management framework.

______ Review the three largest projects you executed last year. Consider how you might have managed them differently with regard to the KAs discussed in this chapter.

Chapter Five: Project Management Phases and Tasks

In this chapter we'll look at some of the "skeleton" of your project so you see what the component pieces are. Starting in the next chapter we'll look at the nuts-and-bolts of actually running a project in the small business environment.

Phases – Process Groups

Any project must be broken down into bite-sized pieces for several reasons, not the least of which is manageability. That's why we call it Project Management. One of the ways we break down a project is called process groups (or phases). Just about every project has these groups in common.

We will use the term "phase" instead of process group because it's easier and fits nicely with our way of running project.

A typical project's phases follow a logical progression. This format is called a "Waterfall Methodology" because each phase of the project feeds into or "falls" into the next part. For the most part (but not always) they are dependent on the completion of the previous phase.

And because these phases are sequential, they lend themselves to convenient tracking on any Project Management tool, the most common of which is Microsoft Project. The authors highly recommend that *every* project – no matter how small – be tracked in some fashion. Ideally, you'll use a PSA and ticketing system. This is the surest way to guarantee that you remain profitable on every project.

1. Initiation

First we think about our project and get some rudimentary ideas on what we want the project to do, a ballpark cost estimate, and some other things. We call this "Project Initiation."

2. Planning

During Planning, we might talk to our client, or their accountant, or some key stakeholders (users) and get their feedback. We might work the numbers in greater detail, decide if the outcome is worth the resources (Cost-Benefit Analysis), we might plan a basic timeline, make some phone calls and get pricing ... maybe even a hard quote on equipment, etc.

If things look good, and we decide to go forward, this is the time we'll make our plans. We will document the desired outcome (the deliverables), how long things will take, and anything else we need to do before we do any actual work. We call this "Project Planning." This is where we actually do most or all of the planning necessary to make our project successful.

This is also the time when we gather our project team and brainstorm about anything that might go wrong once we start the project work. The Planning phase is the time we do any and all of the pre-project planning necessary to make our project successful, so if your project has specific planning tasks that make sense, this is the time to get them started.

3. Execution

The next phase is called "Execution." This is where we do the actual work of our project: install the server or workstations, upgrade the OS or application, run the T1 from point-A to point-B, or whatever we had planned on accomplishing. If we have a software project (SDLC – or Software Development Life Cycle), this is where we deploy it after our final testing, often a "user acceptance testing" effort.

When we finish the Execution phase, we are almost done with our project. But we still have a few loose ends to tie up before we go celebrate.

4. Closing

The final phase is called "Project Closeout," or the "Closing." The is where we tie up those loose ends, make sure everything works, maybe if we're diligent, we might write up the things that challenged us and how we took care of them … this is called "Lessons Learned," so that if and when we do this project again, we have some information to help us speed things along. This is also the time we get our key sponsor and/or stakeholder to sign off (if there is a formal document).

Last but not least, we all go celebrate our wins. This is very important, and many organizations brush this off as unnecessary.

Team celebrations have many benefits, not all quantifiable or realized immediately: they build team unity for future endeavors, and they also let everyone take a well-earned break before the next project gets going. Pay attention bosses: your team needs a breather.

5. Monitoring and Control

The PMBOK Guide also lists a 5th process called Monitoring and Control, which occurs throughout the entire life cycle of the project. Most of the remainder of this book is dedicated to monitoring and control. We propose a specific way to do this successfully in small business I.T. projects.

Agile Project Management

A different methodology has emerged called "Agile" methodology where the project work is broken down further into bite-sized pieces. Each piece is performed as a whole with corrections or changes being incorporated on the fly. Constant measurements provide real-time knowledge of resource expenditures as well as project accomplishments. The certification is called a CSM, or Certified Scrum Master after the rugby game where small movements toward a larger goal are performed.

Project Management Tasks

This section provides a simple set of checklists for you to use as you work through the phases of your project. First however, you should read or review the Knowledge Area (KA) chart in Chapter Four and become familiar with the various (compartmentalized) areas you'll be undertaking. Decide in your mind which areas will require the most of your and your team's time and energy. It's highly advisable that you make some notes before continuing with the checklist and as you read through it.

The first thing to do however is to assess the size of your organization and choose which of the KA's you'll be using. You might decide to skip one or two KA's for the sake of expediency or budget. You might also decide to place heavier emphasis on one (Risk for example) to bump up your probability of success. This is especially true if there are unknowns in your project, or if this area of endeavor is new to you. YOU are the Project Manager (PM). You may well be the engineer, developer, CEO and President of the company, but for now you are driving this project forward and will be wearing the Project Manager's hat.

This checklist is broken down by the four main sequential process groups that make up the majority of small IT projects. As mentioned in the introduction, they're chronological, and sequential. Each group is dependent on the prior group (Execution can't happen until Planning is complete), so feel free to chart the progress on a spreadsheet or a Gantt chart (MS Project for example). A whiteboard works well too ☺.

This list is a summary. The detailed checklist follows.

A. Initiation:

   a. Charter / SOW (statement of work) / Scope doc
   - Get this signed, list deliverables, time frame, budget

b. Create stakeholder list

B. Planning:

a. Document actual work to be done and define deliverables

b. Do vendor research, get quotes, obtain vendor agreements

c. Gather team

d. Develop risk plan

e. Develop preliminary schedule and decide how it will be tracked

f. Develop preliminary budget (+/- 25%) and decide how it will be tracked and managed (and communicated)

g. Document stakeholder and/or communication plan

C. Execution and Monitoring/Controlling (combined for small projects)

a. Do the actual work

b. Ongoing schedule tracking and management

c. Ongoing management of the scope and deliverables

d. Ongoing reporting (communications) and stakeholder management

e. Ongoing budget tracking and management

f. Ongoing team management

g. Ongoing risk management (tracking is optional)

D. Closing

a. Document lessons learned

b. Obtain final approval and/or signoff of all contracts

c. Make sure vendors/invoices were paid

d. Celebrate

And now the detailed checklist:

A. Initiation:

- ☐ Draft your SOW/Scope document:
    - Decide deliverables, time frame, preliminary budget (+/- 25%), type of agreement (time and materials, fixed price, etc.)

B. Planning:

- ☐ Finalize your SOW/Scope document
  - This is the most important single thing you can do on your project. Extra time and effort here will be well worth it.

- ☐ Execute (both parties sign and date) the above document and distribute copies
- ☐ Create a stakeholder list
- ☐ Create a working project document for you and your team (the actual work to be done and the resulting deliverables)
- ☐ Do vendor research, get quotes, obtain vendor agreements
- ☐ Gather your project team
- ☐ Create and execute any relationship documents for your project team
- ☐ Develop a risk plan and create the working documents
- ☐ Develop a preliminary schedule and decide how it will be tracked
- ☐ Develop preliminary budget and decide how it will be tracked
- ☐ Create schedule and budget tracking documents
- ☐ Create stakeholder and/or communication plan and document

C. Execution and Monitoring/Controlling

- ☐ Do the actual work
- ☐ Track the schedule
- ☐ Track the budget
- ☐ Track the quality of the deliverables
- ☐ Communicate as per your plan (manage your stakeholders)
- ☐ Manage your team
- ☐ Manage your risks and issues

D. Closing

- ☐ Obtain final approval and/or signoff of all contracts
- ☐ Document any lessons learned
- ☐ Make sure vendors/invoices were paid
- ☐ Celebrate ☺

Okay. Are you ready for the nitty gritty?

The remainder of this book presents a series of forms to build your project binders, along with descriptions, explanations, and examples. We cover the literal click-by-click of project management in the Small Business environment.

In the next few chapters we walk through some real-world projects that you are probably already doing, and we show you how they would look using the forms we provide in Appendix A.

Notes:

Section 3 - The Project Binder Process

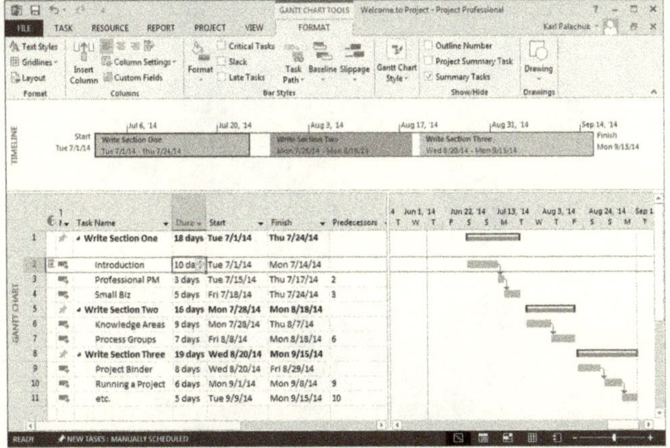

Chapter Six:
Introducing the Project Binder

In this chapter we're going to look at the forms that make up the Project Binder. We'll have a quick overview so you get a sense of how the process works. In Chapter Eight we'll discuss how we recommend using project binders and discuss project management generally.

In *Appendix A* you'll find a copy of all the forms included in this kit. You can download these forms as part of the downloadable content once you register this book. See the instructions at the beginning of the book. For now, put your thumb in Appendix A so we can go over it together.

Note that you can customize this document as needed.* For now, just print it out as is.

* While you can use any paper you wish, we recommend that all documents with the KPEnterprises logo be printed on fine linen paper with a watermark. ☺

Here are the forms you'll find:

Project Title Page

Very straight forward. Put in a descriptive title. We like to slip this page into a "view" binder. That's the kind where you can slip a printed page into a clear pocket on the front of the binder. If you have several projects for one client, you might create one binder and put tabs between projects.

The project will live in this binder until it is complete. At that time, you can remove all the pages, staple them together, and file them away. This is more convenient than keeping every project permanently stored in a binder.

Status Page

This page will give you the last action item and the next action item at any time in the project – if you keep it up to date. This is one of the key tools that will allow you to put down a project and pick it up again.

Similarly, this page will allow the project manager to flip open the binder and know whether anything has progressed since the last time he reviewed it.

Table of Contents

This is pretty standard stuff. Just a TOC page.

1. Project Overview

In most cases, this is a sentence or two. After you get used to the process of creating projects, the text here will be essentially identical to the service ticket title and description.

Key to Your Success

If you recoil at the prospect of writing 1-2 sentences to describe the project, that's probably a sign that you're not ready to start using project binders. Perhaps you're not committed to the process.

Our plea is this: Don't walk away. Finish reading this guide. Commit to the process. Work your way through a project.

We firmly believe that standardized procedures such as this will make your business more profitable. We'll cover this more thoroughly in the next section. But for now, please believe us that you need to do this.

2. Project Goals

The goals of your project describe what you're trying to accomplish. Unlike more objective "action items," the goals are intended to address a bit of the big picture.

We always try to take a step back and ask about the big picture when a client asks us to do something. Sometimes a client asks for something because, based on their knowledge, that thing fulfills a need. But when we ask what the big picture is, we find that there's a much better solution for them. They just didn't know to ask.

Similarly, when we look at the big picture we might decide to drop a project altogether, or change it significantly. For example, when we replace Sue's computer, we will re-evaluate older service tickets to troubleshoot errors, update software, or add more memory in her old machine.

The other, more obvious reason for writing down our goals is to set a benchmark for success. This is particularly important if this project is a small piece of a larger project.

Anyway, write down one or two simple goals for your project. Then, when it comes time to define actions that will lead to goals, you'll be able to address the question "How does *this* action help us reach *that* goal?"

Here's a great tip for success: Write these goals so your client can understand them. Verify that the *client can explain to you* what these goals are – in people-speak not tech talk.

3. Current State and **4. Desired State**

What is the current state of the system? Describe the status quo. This will help you describe the next section, which is what the system will look like when you've successfully completed the project.

You may wish to simply describe the current status, or you may use diagrams if they help. For example, if you're moving a server from the LAN to a DMZ, before and after diagrams may be very useful.

For technicians who are overwhelmed by the prospect of creating a project, these before and after descriptions can be very helpful. After all, the next task is to describe how you get from where you are to where you want to be.

The goals should be written so the client can understand. Here, the Current State and Desired State should be written so that any technician who reads them will understand. This can be as techie as it needs to be.

5. Stages of the Process

Finally we get to the *action steps*. Each stage follows a set pattern that includes:

5.1. Discovery - Documenting pre-existing network or other environment

5.2. Define training needed by client or technicians

5.3. Define the Timeline

5.4. Stage _____1_____ _____2_____ _____3_____ etc.

5.4.1. Actions

5.4.2. Notes

5.4.3. Additional work/issues to "Additional Work" Stage

Note: Repeat the stages as needed until all work is defined. There will be as many stages as there needs to be. This means you'll print out or photocopy the "stage" forms as many times as needed.

5.5. Fine-Tuning to make sure all stages are complete and all goals accomplished

5.6. New System documented

5.7. Additional Work Stage

All "additional" work that comes up flows to this stage.

5.8. Deliver Client Training if Needed

5.9. Project completed and signed off

Each stage has a definitive ending or completion. All work that's un-done or needs to be redone flows into the "additional work" section.

When each stage is completed and all additional work has been entered into a ticketing system, or to-do list, then the project is complete.

Key to Your Success

Perhaps the biggest reason that projects go over-budget is that projects *creep*. You know how it is: You go in to install a firewall. You get side-tracked with configuring a spam device, a terminal server, and a printer. The client says you promised two hours.

Your super power is the Service Request.

"I'm here to install this firewall. For the other items, either you can create a service request or we can create a service request."

When each task has a service request, you can be sure each thing gets done. And if something goes wrong, you can keep track of the hours and make sure you've isolated all work on that

SR. You might not end up billing for all the work, but you make sure that your work on other service requests remains billable.

On a more macro scale, Projects do the same thing. There's a Scope of Work. So you'll discuss the scope of work with your client. And once this idea is introduced, you'll be able to discuss topics as being *inside* the scope of work and *outside* the scope of work.

6. Final Check-off for Project Plan

Remember: At this point you haven't *done* anything. You've just laid out the plan for how you will do something. Once your project plan is complete, you'll take it to the client and make sure they agree to the plan.

If you haven't done so before, this is where you'll discuss costs and project variables. There might be a bit of sticker shock. But you'll also be in a great position to make changes based on a realistic assessment of what it will take to get the job done.

Again, here's a bit of reality for the SMB consultant. Most SMB clients don't care about "your" project. But it's important that you go through the key points anyway. This allows you make sure the client knows the costs, the expected schedule, and the expected downtime.

Most SMB consultants admit that they're really bad about forcing their clients to sit down and go over this kind of thing. And what are the clients' biggest complaints about projects? Unexpected costs. Unexpected downtime. Poor communications.

7. Project Evaluation

This isn't a form, really. It's just a place to start taking some notes. When your project is complete, it is critical that you learn what went particularly well, what went wrong, and how you can improve the whole process.

Concluding Thoughts on the Binder Forms

Please consider these forms as a starting place. Your organization might be a bit different. Your terminology might be different. With luck, you'll add some things and fine-tune your process.

Here's what we recommend for the Project Binder document. Keep it where it's easy to find. Maybe print off a few copies and have them ready to go. Whenever you have a project on the horizon, build a binder.

As you go through each project, keep notes for yourself and consciously fine-tune the process. As the process evolves to reflect your business, you'll have even more success with project binders. And as you learn the binder project process, your business habits will also improve.

Your To-Do List for The Chapter

_______ Print off a full set of forms

_______ Customize the forms as you see fit (logos, headers, footers). Store your template forms in a location accessible to your team.

_______ Build at least one Binder with all the forms

Chapter Seven:
"Running" a Project with the Binder Process

Here's where we are: We've done an introduction to the practical side of project management for SMB. We've examined the *knowledge areas*, *process groups*, and *tasks*. You have the forms to create a binder. With luck, you've already printed them out and put them into a binder.

This chapter will address the bigger picture of running your projects. Karl has written a few "forms" books, so he's noticed that people have the tendency to print out the forms and not read the rest. Or, worse, flip through it once and then not look at it again.

But we believe that this simple project management system can improve your business – and your profitability. Here's an overview of how a project works:

- Start Project
    - Start Binder
        - Define Project (form)
        - Execute Project
        - -- Stages, Notes, Timelines, Statuses, etc.
        - Complete Project Stages
        - Complete Paperwork / Get Signoff
        - Complete Evaluation
    - Finish up Binder
- Complete Project

Please note: The forms in your project binder are the middle part of this. And Section 5, the "Stages" or action steps, are the core of the core. The forms help you organize that part of the project so that everything gets done and flows as it should.

But the overall project begins before you print the forms, and finishes only after all the forms are filed away in a file cabinet.

So, let's start all over and look at real-world project management in the SMB space, using the Project Binder described in this book.

Real World SMB Project Management

Real world SMB project management starts with a simple request, or a need identified by the consultant. We'll use, for our example, the movement of a web site from an in-house server to a hosted server. Here's the background:

You have a new client. They've had a simple five-page web site for many years. And it looks like it. They have hired Cousin Larry's Pretty Good Web Site Design to build a new site, and to host it. You've been asked to help facilitate this move. The client doesn't want to have a meeting about it. "You and Cousin Larry just take care of it."

So you find yourself with a pretty clear endpoint, no real direction on how to get there (because that's your job), and no budget to speak of.

Your tendency might be to call Cousin Larry and get the ball rolling.

No.

You don't have a service request and you don't have a project binder.

Nay Sayers Say: "Hey, Karl, you're insane. This is really simple. We do this all the time. Just do it."

As you're driving back to your office, you get a call from the client. "Did you do something while you were here? Our email isn't working. We can get to the Internet, but we can't get any email. Our customers say email to us is bouncing."

Okay. Now you start a service request. You need to be "on the clock" because you have to call Cousin Larry and ask whether he took control of DNS. Yes. Of course. They always do that. And the MX records? Well, they point to the new web server, of course. And there isn't even a web site there yet! The client doesn't know that his web site's down because he doesn't browse his own web site.

You explain to Cousin Larry that he has messed up the client's email. And wouldn't it be better if the WWW host were pointed to the old site until a new site exists? He explains that this has never happened before. You convince him to point the records back to where they were and explain that:

1) You will control DNS. Period.

2) He will tell you the new address for the web site and you'll point to it when it's time.

By the way, how did he change the DNS records? The client gave him his Network Solutions password.

You create a Service Request to gain access to DNS as the technical contact and manage this for the client. This is outside the scope of the project and billed separately. You begin working this SR.

Now, we hope, you see that even simple tasks are projects and need to be treated as such.

So we're back to square one. Let's start this project.

New Project, Step One – Create SR(s)

The first thing you need to do is to decide whether this project will use one service request or more than one service request. If you have a simple project, it often makes sense to use a single service request. You will simply list major stages in the SR and keep the details in the project binder.

If you have a really serious project, such as a network migration, then you'll want several SRs. In this case, we think it does make sense to have a single SR. We'll call the SR "Move web site to Cousin Larry's Hosting."

Click. Enter. SR created. Now you look at your watch and start billing time against it.

Project, Step Two – Prep the Binder

Print out a set of forms, 3-hole punch them, and put them in the binder. Make a label for the spine. Fill out the first sheet (name of project, client, project manager, project leader). Do you have a due date? If not, go to the status page and make an entry under "Next Action Step" to determine due date.

Proceed to fill out the forms to the best of your ability. Very soon we'll print out the forms from this and other sample projects. Here we'll just describe the project and not fill out the forms for you.

Project Description. This is an overview. 1-2 sentences on what the client wants. You're keeping email in-house and just moving the web site out of house. Primary goals should include zero downtime for any services.

Definition of Goals. These are visible action steps. For example, determine future IP address of web server. Make sure domain administration is secured so Cousin Larry can't mess it up again. Update DNS. Decommission old web site.

Define Status Quo. Either via network documentation forms or Visio diagrams, or both, describe where all relevant servers and services are now. For example, see Diagram 7-1.

Define Where You Want to Be. Either via network documentation forms or Visio diagrams, or both, describe where all relevant servers and services will be. For example, see Diagram 7-2.

Internet
Web and Email Server
- Web Site
- Email
- OWA
Viewed on the Internet as
www.domain.com
www.domain.com/exchange

Diagram 7-1

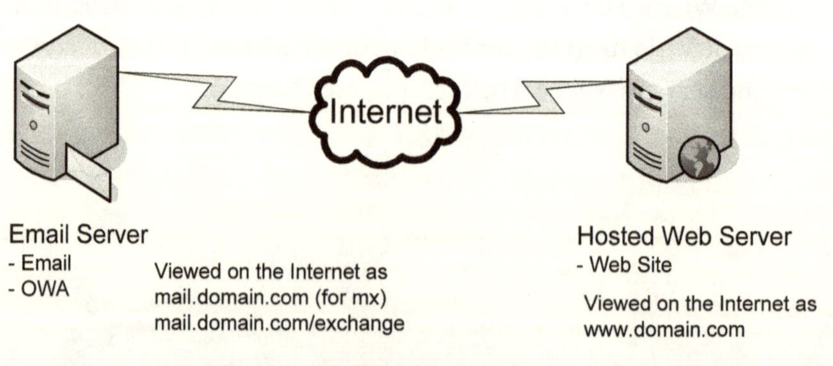
Internet
Email Server
- Email
- OWA
Viewed on the Internet as
mail.domain.com (for mx)
mail.domain.com/exchange
Hosted Web Server
- Web Site
Viewed on the Internet as
www.domain.com

Diagram 7-2

We'll do a "Discovery" process, but right now it looks like one of our steps is going to be to change the machine name to the outside world. That means we have to create mail.domain.com and point mx records to that. No problem. Make a note. Check to see whether there's a secure certificate, or if it's used. A new certificate will probably have to be generated.

Project, Step Three – Define The Work

Step Three is to lay out the stages. If you think you know what needs to be done first, second, third, then start filling out the pages. You don't even have to use full sentences, but you need something on every single page.

Because this is a new client, let's not fill out all the stages until we do the discovery process.
So you go to do your discovery process. You sit down at the server and get past the screen saver. You guessed it:

Small Business Server 2011

Okay, then. That changes everything.
And, in a sense, it changes nothing. This project was going to involve RWA and SharePoint whether you guessed that or not. The only thing that's changed is your knowledge of the situation.

So you do a thorough review of the server, and what it's being used for. You interview the client. Are they using VPN? No. You back up the firewall and then examine all of its mappings.

Key to Your Success

The **Discovery Process** is one of the most important, and most neglected of all stages in a successful project.

One of the lessons that is beaten into your skull endlessly in formal project management training is to make sure you know what you're getting into. And, unfortunately, it is one of the steps most commonly skipped by SMB Consultants, often at the insistence of the client!

You're done with that now.

From now on, you inform the client that you're going to do a complete discovery. Yes it's billable, but your experience is that it always saves money in the long run.

By now you've gained access to DNS. You print out all settings and records.

Once the discovery process is completed, you have revisions for the before and after diagrams. Diagram 7-1R is the revised version reflecting the SBS setup.

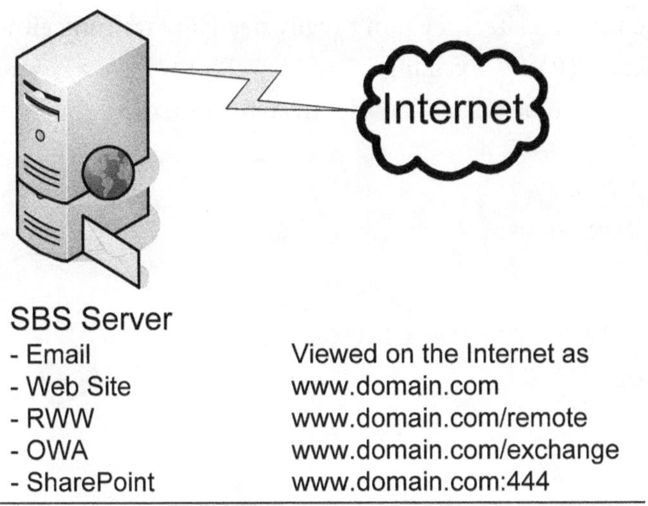

Diagram 7-1R (revised)

And, of course, your "After" diagram will also be different. You've decided not to use the fully qualified domain name (FQDN) of mail.domain.com and opted for sbs.domain.com. In either case, www needs to point to the hosting company. For completeness, you verify that no https is being used for the existing web site except for the special features of SBS. You also verify that the client's email is not going through a spam filtering service that uses DNS to find servers. So no changes are needed there.

So now you have an excellent picture of where you are and where you want to be.

Next, you proceed to lay out the training that will be needed. No one on your staff needs training. And since the client does not manage the web site, they don't really need any training either. Remote access (RWA, Exchange, and SharePoint) have a new address, but you should be able to handle this with a memo.

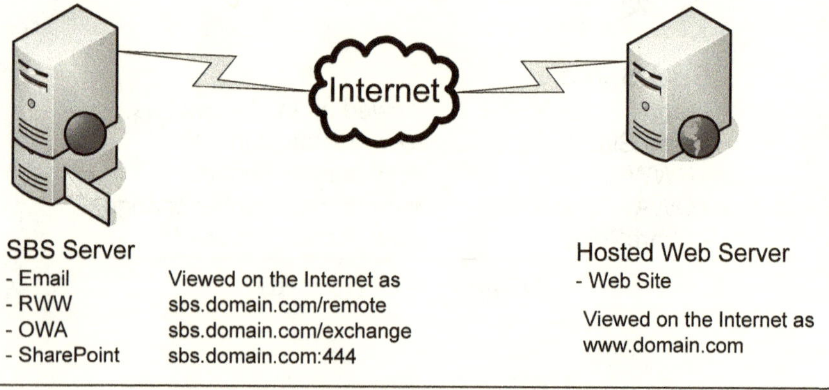

Diagram 7-2R (revised)

You decide to train one key person on the change from *www* to *sbs*. That person will train everyone else.

After discovery, you are now able to estimate the time more accurately and discuss this with the client. You contact Cousin Larry to determine when the new web site will be ready. He says two weeks. With the client, you set a date to have all of your work done within one week. Then, when the new web site's ready, all you have to do is to change one DNS entry.

You define the following stages:

1. Create DNS entry for sbs.domain.com.
2. Change mx records from www.domain.com to sbs.domain.com.
3. Change secure certificate so that it reflects the FQDN sbs.domain.com.
4. Test. Test Everything. Test that email flows in and out with new mx records. Test RWA with new certificate. Test OWA with new certificate. Test SharePoint with new certificate. Configure an Outlook client and test Outlook connection with RPC over HTTPS.
5. Update any client how-to documentation re: Outlook, SharePoint, OWA, and RWA. If it doesn't exist, create it.
6. Identify all users who access Outlook using RPC over https. They will need the new certificate and a configuration change.
7. Identify all users who access SharePoint. They will need to know to install the new certificate.
8. Inform Cousin Larry's Pretty Good Hosting Service, in writing, that they need to stop any DNS hosting that they may have created for this client. Anything they have could make a mess, even if it's only for their own customers.
9. After new web site is up, change DNS to point www to new IP.
10. Document everything in the Network Documentation Binder.

Reorganize

It's very important that you take a minute and review the stages. Your first goal here is to completely list every single step that's needed. So this could be a bit of a brain storm. As a result, some stages may be out of order. Let's do a little rework.

Perhaps:

1. Create DNS entry for sbs.domain.com.
2. Change mx records from www.domain.com to sbs.domain.com.
3. Inform Cousin Larry's Pretty Good Hosting Service, in writing, that they need to stop any DNS hosting that they may have created for this client. Anything they have could make a mess, even if it's only for their own customers.
4. Identify all users who access Outlook using RPC over https. They will need the new certificate and a configuration change. Inform them that a change is coming. Identify users who will need help.
5. Identify all user who access SharePoint. They will need to know to install the new certificate. Inform them that a change is coming. Identify users who will need help.
6. Update any client how-to documentation re: Outlook, SharePoint, OWA, and RWW. If it doesn't exist, create it.
7. Change secure certificate so that it reflects the FQDN sbs.domain.com.

8. Test. Test Everything. Test that email flows in and out with new mx records. Test RWW with new certificate. Test OWA with new certificate. Test SharePoint with new certificate. Configure an Outlook client and test Outlook connection with RPC over HTTPS.
9. For all clients who need assistance with the new configurations, help them get new certificate installed and working.
10. After new web site is up, change DNS to point www to new IP.
11. Document everything in the Network Documentation Binder.

Here's what we changed: After reviewing the sequence, we decided to make sure we gave adequate notice and information about the change. This led to identifying users who would need help with the new certificate. That led to an additional stage to help these people out.

We want you to look back for a minute to the first page in this chapter. Note, please, that we haven't done any work on this project yet, except to define these stages. Right now we're in the middle of making sure we know where we're going and how we're going to get there.

We're at the part that says "Define Project." Once we have all those forms filled out, we're going to get it approved. What does that mean?

Well, if the client will have a conversation, then you take it to the client and say "Here's our plan." Very often in the SMB space, the client will simply say that you should just do what you think is right.

If you're a multi-person shop, then someone needs to look at this and agree that it's set up right. That might be your manager, or a technician who works for you. You just need a second set of technical eyes.

Note on Time and Money: Too often, clients are surprised because a simple little thing turned out to be expensive. At the beginning of this we played a little devil's advocate and addressed whether this was really worthy of a project. After all, it's so simple.

Well, look at it. Let's say this client has ten users. Three of them need assistance with Outlook and certificates. Let's do some time estimates for our eleven stages. Yours may vary, but substitute yours for ours.

Defining Project, managing Project
.50 hours

Stage One: Create DNS entry

And . . .

Stage Two: Change mx records

And

Stage Three: Email to Cousin Larry
Combined .25 hours

Stage Four: Identify users who access Outlook using RPC over https

And

Stage Five: Identify users who access SharePoint
Combined .50 hours

Stage Six: Update documentation
.50 hours

Stage Seven: Change secure certificate

.25 hours

Stage Eight: Test Everything

.25 hours

Stage Nine: Help clients with new configuration 1.00 hours

Stage Ten: Point DNS to new www

.25 hours

Stage Eleven: Document everything

.25 hours

Grand Total: 3.75 hours

So that's about $450 if you charge $120/hr. And that assumes that all goes smoothly and you don't spend time standing around waiting for someone or something. It would be very reasonable to estimate five hours for this job.

At a minimum, you need to have a discussion with the client that this project will take some time. Assure him that they won't have any downtime. But it will cost money to get this done. And since you have a project binder under your arm, you can discuss it to whatever length he wants.

Again, our experience is that the client won't want to discuss it, and he will want you to just do it. But at least now he knows that it's more complicated than changing a DNS entry. In fact, just changing a DNS entry put his email offline for a couple of hours. You won't let that happen again.

Key to Your Success

The Status Page.

At the beginning of the binder is a status page. This is the only page that lists actions taken and next action step. This little page, if used, will allow you to know exactly where you are in the project.

Once you complete writing out all the stages, note that. Your next action step is to begin the project! Yeah. As you begin Stage One, note that. When you complete Stage One, note that. Make it your new habit of success.

Project, Step Four – Execution

Now we get to begin the work!

As you enter the "Execute" phase, you will step through the first stage. Make a note on the status page, and begin Stage One.

Do the work. Take notes.

Remember that every stage has a place to note where things aren't quite right, or where additional work crops up. For example, in the middle of this you find two users whose email is having problems. Write it down. Make sure you don't forget it. But don't work on that right now. Another user wants to upload a whole folder full of files to SharePoint. Make a note. Don't do it now.

When you're finished with each stage, copy all these items to the Fine-Tuning section. You can create individual service requests now or at the end of the project, depending on what makes the most sense. Remember, this is one of the keys to your success in any project.

Fight Project Creep!

Practice these words: "That's outside the scope of this project."

Project, Step Five – Finish Up

When all the stages are complete, and you've tested everything, and the client's been trained, then the project is "complete." You're not done yet, but the project is complete.

It's very important that all that spin-off and extra work get into new service requests. It's equally important that the documentation is up to date.

Now we're "outside" the binder. When everything in the project, and in the project binder, is complete, then you need to do an evaluation. Again, this is an important step that is frequently skipped. We're too busy. We have too many projects going. No one's motivated. Excuses, excuses.

Your project evaluation might take all of ten minutes. But it needs to be done. What went right? What went wrong? Is this process clean enough to take to another client and do again?

In our example, this process is pretty clean. You can easily see how this project could be made just a little generic. Maybe put the DNS configuration inside the project instead of out.

And there you have a procedure to move an in-house web site to an out-of-house hosting service.

You can do this again and again. And make money every time. You can literally prepackage your "Web Site Move" project and only need to make minor changes per client.

Anyway, the project is not complete until you check through every page. Fill out everything that didn't get filled out. Document everything. Do the evaluation.

Then you can take the pages out of the binder and put them in a folder.

Key to Your Success

Never throw away a project. Take it out of the binder and put it in a folder. File by client, project title, and date.

If someone asks the question "How did we get to this point?" you will have an answer!

In addition, you never know when you'll need to do a similar project. The project binder is a resource during the project, and a potential resource after the project. At a minimum, it's a record of what you did.

Label the Project

Create a label to place on a file folder or file pouch. It should include the client's name, the project title, and the date, like this:

ABC Co. – SQL Migration – 2015-Jan

You might also want to add a few notes on the outside of the folder that might be useful after everyone has forgotten exactly what you did on this job, such as:

✓ Additional work: Move email from in-house to cloud

Your To-Do List for The Chapter

______ Print out the binder forms and enter in the stages for this project

______ Think about two of your "representative" clients and customize this project for each of them (e.g., a 5-user with a local ISP and a 25-user with cable Internet)

Notes

Section 4 - Tracking the Business Side of Project Management

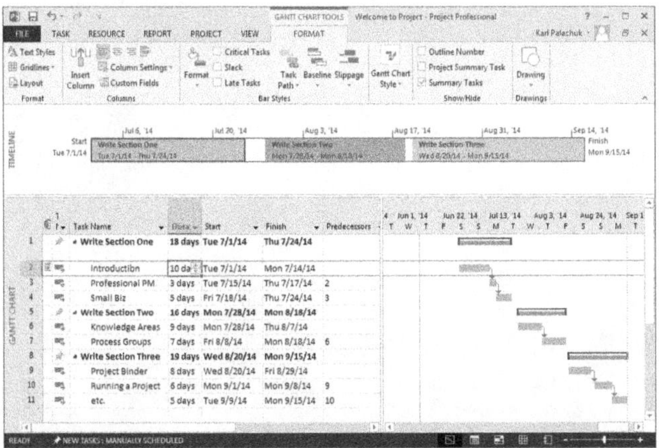

Chapter Eight: Quoting the Project Correctly

Quoting the Migration Project

This section covers the most important aspects of packaging, estimating, and quoting the migration project. Note: This is not a sales primer, so we don't cover the fine points of getting the client to sign on the dotted line. For that, we refer you to the excellent sales books by Erick Simpson and Matt Makowicz. For more information, see Appendix C: Resources.

Our goal here is to cover the total management of the project. So, this chapter will address the sales process only insomuch as it affects the management of the project as a whole.

Very frequently, project definition and delivery involve three roles within your company:

- Sales
- Sales engineer
- Engineer

The sales role focuses on pricing, packaging, and client agreement. The sales engineer role is responsible for making sure the right thing is sold and making sure it can be delivered by your company. The engineer role will actually deliver the network migration project.

For most companies, these roles are played by at least two or three people. In larger projects, a team will deliver the engineering role. But it is also common for these roles to be combined. If you're a sole proprietor, you'll deliver all three.

It's very important that you do deliver all these roles. Do not skimp on any of these roles. Selling the right project is just as important as delivering the project that's sold.

Project Initiation

For the purposes of this chapter, we're going to look at a larger project. We'll use a new server installation and migration as the example.

At some point, you'll find yourself with the opportunity to propose a new server. For the purposes of this chapter, we'll assume you've had an initial meeting, you've discussed the new system with the client, and you've been asked to put together a quote.

In a perfect world, you'll be able to charge for a complete audit of the client's systems. But, as we'll see, that's unlikely. It is more likely that you'll need to leave this meeting with permission to poke around on the server and ask someone a few questions.

You should try to learn as much as you can about the server, particularly the amount of hard disc space in use, any databases that exist, and any LOBs (Line of Business applications).

If you're maintaining this site for a client, your job will be much easier. If not, try to get the basic information on our "68point checklist" for examining client systems. You can get an electronic copy of this list for free at SMBBooks.com. Just look in the "Free Stuff" category

A Confident Approach

It is important, for your long term success that you take charge and control the entire project as early as possible. Here's what we mean by that: technology users and owners are rarely qualified to design or architect a modern computer network. They might know a great deal. And they might appear to know more than they actually do.

But, assuming you are a qualified technician, you need to take charge and run the show. The reasons for this are fairly obvious: You know how to architect an entire network; you know which hardware and software options are available; you know what's best for the job; and you know which alternatives are available while discussing options, expandability, etc.

In other words, you understand the technology. You can try to explain it to the client, but in the end they will never understand it as you do.

Let us put it this way: You're the consultant! It's your job to consult!

Your job is to help the client define the right network for their needs. Never simply give the client what they ask for. In the last twenty years, we have almost never seen a client request for a major network overhaul that turned out to be the right thing for the client. Sometimes, they want to overbuy equipment, sometimes they want to under-buy software. They are rarely aware of the hardware or software vendors' latest packages or recommendations.

There are essentially three ways that large technology projects are initiated:

- By the client
- By the consultant
- By God (or "an act of God")

From time to time a client will simply come to you and say "That server's getting old." At other times, you need to go to them and remind them that the server's getting old, or that it's not fulfilling their needs. And finally, you have times when something fails and you find yourself patching things together long enough to get a replacement set up.

Luckily, that third option is rare these days.

No matter how the discussion starts, it's your job to move quickly to guide the conversation. In addition to technical knowledge, you will also rely on the client's needs, their plans for growth, their budget, etc.

You might be wondering why it's so important to take the project by the horns and lead the conversation from the start. The answer is pretty simple: you're planning your success. It's usually surprising how much you can control success when you control the conversation. Here's why.

There are two key factors that affect your relationship with business decision makers: their level of confidence in you, and their level of technical knowledge.

If their confidence in you is high, then you have no problems. But they can lose confidence if you're not a take-charge person. And, as you can imagine, their confidence will go way up as you assume control of the conversation.

Technical confidence is a little different. If their technical knowledge is low, they really need you to take charge. They actually need to be told what to do. If you come across as uncertain, you make it more difficult for them to agree to your proposal. If you give them two options, you'll need to state a preference for one over the other.

A high level of confidence on your part makes it easier for them to go along. In fact, you'll hear the client echoing your arguments as if they're facts. "Of course we need RAID 5 on the storage array."

If the client's technical knowledge is high, their tendency to challenge your recommendation is also high. You'll need good, clear explanations for your recommendations, but your sense of confidence will win the day.

Remember, also, that what appears as a challenge may simply be the client showing off a level of knowledge. If the client takes pride in knowing about technology, then they want to have a true discussion about the options available for their new network.

And best of all, once educated, the client is more likely to go along with your recommendations.

We have a client who owns two different companies. In discussing the benefits for a hardware-based RAID system for his new SBS 2008 installation, he really latched onto the idea of gaining speed with RAID five because the system was pulling data from more heads at the same time.

Literally one month, later he asked us for a quote to fill up the one-year-old server at his old office with hard drives. We told him this was not necessary, but he was eager to spend the money. What can you do?

We tend to only make one set of recommendations for a prospect: the one we consider to be the best choice for the client. Unless the client really pushes, we don't give them choices. We'd much rather spend time learning about what the client needs than arguing with them about what they need. In fact, when we propose a new system for a client with a very high level of technical competence, we actually tell them "this is the only option we're quoting because it's the right system for you."

The bottom line is: Have confidence in what you propose and propose it in a confident manner.

The Generic Two-Drawing Process

We like to give the client two basic Visio diagrams to work with. The first is a drawing of their current network—what is. The second is our proposed network—what it will be.

Drawings are great because you can take the client on a tour from the firewall to the desktop, and everything in between. Is the firewall adequate? Do you need to add web filtering, VPN, etc.? Next are your switches. Do you have good gigabit switches that are still under warranty?

Move like this through the entire network. As a rule, you'll address one thing at a time. But, when you make your proposal, the client will know that you really understand their environment.

As you can see, this entire approach is focused on what's best for the client. Yes, it requires a certain ego. But when you confidently propose only what's right and best for the client, the client will see that.

When you sit down with these maps, start by describing the first map strictly in terms of functionality. For example,

"Currently, your mail lives at the ISP. That means two important things for you. First, when you send a file to Jennifer, it has to leave your company and sit on the ISP's computer. Then, when she collects her email, she brings it back in-house. That's very inefficient.

Second, you don't currently have the ability to share calendars or use the other features of Microsoft Exchange Server. Let me show you how that works . . .."

You should go through the entire "old" map first. We recommend that you do not flip back and forth between the two maps. That's a strong tendency for some technical people. Unless the client initiates this back-and-forth activity, you should avoid it because it can be very confusing. If the client initiates, they may be very visual and soak up the information more quickly that way.

Go through the entire map. The old system has no firewall. The old system has an old, slow (often full) switch. The old system has an old Small Business Server.

Once you get to the new map, it's like a shiny new toy. Suddenly they get a firewall, a gigabit switch, hosted Exchange email, off-site backup, etc.

And while every network is unique like a snowflake, the truth is that these network maps fit 90% of most small businesses for the "before and after" conversation.

We recommend that you go on their web site and snag their logo to paste into Visio. A small touch that they'll appreciate.

The Reality of Quoting in the SMB Space

In large companies, and with large projects, there is a standard process that companies go through. There's an official Request for Quote (RFQ), an official response (formal quote), a discovery process, and generally at least three bids are considered.

But that's not what the small/medium business world is like, at least most of the time. There is rarely an RFQ. And the discovery process can be very expensive, so it is usually cut. In reality, it's not even cut because it never exists. The discovery process is almost never considered as an element in the process.

But the good news is that small businesses don't normally ask three different companies to give them quotes on a project. In most cases, they ask their favorite consultant to give them a quote. If it's really just out of the ball park, they ask for some major fine tuning. But most of the time, they simply require a reasonable justification.

The very good news is that the focus is not on money, per se. In the SMB world, the most important elements in a relationship are communication, trust, and compatible values. So, on one hand, it would be nice to have an official RFQ process and get paid for every element of the process. But, on the other hand, it is nice to know that the focus will be on communication and trust rather than money alone.

It is important to note the fact that small businesses are not likely to pay the money needed to have a thorough discovery process. It means that you cannot estimate a job based on complete knowledge.

Stop: Highlight the previous sentence.

When you get rid of the discovery process and the project plan, then all you can do is to work the project on a "time and materials" basis. But, without any sense of a budget, the client will end up being very unhappy with the cost of the project. Does this sound familiar?

All sales meetings involve a few key people. You get together the four personalities who will define the project:

- The business owner
- The client's technical advisor
- Your sales person
- Your technical advisor

As we all know, these four personalities may be inside two people, or three people, or four people. But in all sales meetings, you will find these four personalities.

KPEnterprises used to do a lot of programming work. Here's a very common scenario from the world of defining a programming project:

- You put together a meeting to design what the project is going to look like. The client says, "Well, we cannot afford the mansion that you have presented. All we can afford is a two room shack. And we can't afford the discovery process. You want too much money just to define this project. So . . . just build a two room shack."
- You dutifully build the two room shack. Then the client says "Uh . . . yeah. There's no kitchen here." And you say "Well, you didn't ask for a kitchen; you asked for a two room shack."
- So you build the kitchen, and they say, "Well, we need a study. And we need a lobby. And we need a front porch. And I told you this was a two story shack, didn't I?" Now the client starts to build what they wanted in the first place.
- But the project is now "time and materials," and they are adding labor and labor and labor. They are hobbling things together and it's not organized. It is not *designed*. It is merely clumped together.

What is that going to lead to? A huge expense, a kludgey project, and an unsatisfied customer.

Unfortunately, this is a very common scenario in small business. No money to plan things properly. So we waste money playing catch up.

Lesson Learned: The core problem in SMB is that we don't get to estimate costs the way big companies do because our clients are not willing to pay much for the estimation part of the process.

So, one of the common things that grew up organically in our SMB space, is that we give away the network evaluation as a way to get our foot in the door. You have basically taken the thing the client needs most (an evaluation of how they should proceed) and you have given it away so that you can get the job.

The result of this is that we get accurate estimates only through experience.

Here's an example from Office 2013.

If you are selling Office 2013, you know about how long that takes on a brand new computer. So, if you have a new laptop and Office 2013, you can say, "I'm going to estimate a half hour per machine." In this scenario, you're safe. But in 2016, when that software is older and has a few service packs, fixes, updates, and patches, it's going to take longer to install. And there is a difference based on the age of the hardware. It might take half an hour, an hour, or even an hour and a half.

As you can see, the accuracy of your estimate changes over time. When a product is new, you have a very low level of accuracy. Think about Windows Server 2012 R2 Essentials. When “Server Essentials” was released, you started out with a very low level of accuracy about how long it would take to get that product installed and configured so that the client can sit down and use it.

And then, as your experience goes up, your accuracy goes up. As the years went by, there were patches, fixes, and updates. With each set of patches, your accuracy went down a bit, then stabilized. When R2 was released, all the patches were rolled up into a new version. So, you very quickly had an accurate estimate of installation time once more.

Overall, the accuracy of your estimates goes up. So, over time, the more stable that product is, and the more experience you have, the more accurate your estimates are.

Software vendors are not always helpful. As you may recall, Microsoft promised that an OEM installation of SBS 2003 would take about 20 minutes! Did anyone actually have that experience? No. Most people planned on 5-10 hours of labor before the user was allowed near the server.

So, our experience is critically important to us. And, obviously, you only get experience by doing something again and again. You can get some server installation experience by reading a book, or by having somebody tell you that they think the install is six or eight hours.

But advice only goes so far. You have to use *your* procedures. And your procedures are different from our procedures. Your checklist is different from our checklist. Your technicians are different from our technicians. So, in the mix of things, you have to figure out how you can deliver that product and have good estimates for labor.

Over time, when your estimates are in that higher accuracy range, you are able to create cookie cutter projects. As you will see, we have a formula for estimating network migrations. We estimate so many hours for each desktop, so many hours for each laptop, so many hours for the server, so many hours for the data, etc.

We have good time estimates for Windows Server 2012 R2 Essentials because it's been out for awhile – and it's R2. But next year

Microsoft will release a new generation of server and our estimates will be very inaccurate due to a total lack of experience. We can guess based on the 2012 version, but that's just a "best" guess.

Back to the Maps

Using the two-network-map process will give you a big step up in making sure that you don't forget anything. As you work your way through the office, you might note that you need a new switch, a firewall, etc.

If you have separate people acting in the roles of sales engineer and sales person, then the sales person will rely on the sales engineer to make up the list of equipment needed and verify that nothing is missed when it's done.

Your Line Card

Whether you call it that or not, you have a standard "Line Card" for your business. That's a list of the standard products and services you sell. In reality, it's the stuff you sell most often. Chances are that you can sell almost anything you want.

But you sell a specific set of solutions. Your set is different from our set. For example, in a nutshell, here's the vast majority of what we sell:

Hardware

- HP Servers
- HP Workstations
- HP Desktops / Laptops

- HP Monitors
- HP Thin Clients
- HP Tape Drives (various)
- HP Printers
- APC UPSs (various)
- Sonicwall Firewalls
- Sonicwall CDP
- Switches (a variety of brand names)

Software

- MS Windows Server (Essentials, Foundation, Standard)
- CALs as needed for all software
- MS SQL Server
- MS Exchange Server
- MS Windows Desktop (versions 7 and 8)
- MS Office (various)
- AVG Anti-Virus
- Diskeeper defragmentation

Materials

- Brand name Cat6 cables
- Brand name tapes (various)
- Misc.

Your mileage will vary. You might sell Cisco firewalls, Dell computers, Symantec Anti-Virus, and a whole host of other products.

The point is: You have a standard set of hardware and software that you offer. The more you know about these things, the more likely you will be able to sell the right solution for a given client.

At some level there's a bit of "When you sell hammers every problem looks like a nail." But realistically, you have a standard map for your SMB clients and a standard solution set. The more you know what you sell, the more profitable you will be.

For example, we also like Watchguard firewalls and the BDR product from Zenith. These are good alternatives to what we have been selling.

But we don't attempt to know or carry every brand of computer hardware or software on Earth. We don't change brands at the drop of a hat.

Keeping our Line Card consistent over time maximizes our relationship with the vendors we choose. It also increases our knowledge of those specific products, including our knowledge of their marketing promotions, rebates, etc.

Know what you know about what you sell.

Labor Estimates

One of the key values you'll need for your internal quoting process (the next step) is the cost of delivering an hour of labor. Like everything else, this isn't as easy as it seems.

The most important thing about this calculation is that it includes a lot more than the employee in question. It takes your whole operation to deliver an hour of labor. You need to look at the big picture rather than a single person.

You need to estimate the fully burdened labor rate. That means the true cost of labor, including taxes, time spent training, drive time, etc. Your fully burdened labor rate is essentially the entire cost of the employee per year divided by the number of hours you'll get.

Note: This is not the place where we figure in the effect of "billable" vs. "non-billable" hours. We take that into consideration when we create quotes. For example, if we think it will take twelve hours of labor to produce eight hours billable, we deal with that on a job-by-job basis. You must be careful not to do this calculation in both places or you'll really burden yourself!

There are some calculators out in the world, but finding just the right one for your business is probably not worthwhile.

Here we present three ways to calculate your cost per hour of labor. There's Super-Fast, Quick, and Complete.

1. *The Super-Fast Method* is this:

Hourly Rate x 1.5

That's it.

Not perfect, but a darn safe place to start. If the technicians working on this project make $20/hour, then your cost is $30/hour.

2. *The Quick Method* is almost as easy. First, you need to know the total cost of running your company. Then, back out costs of goods sold (COGS). This leaves you a total for labor plus all the expenses of running your operation.

Next, you simply divide that by the hours worked by technicians. For example, let's say your cost of operations minus COGS is $200,000 and you have three fulltime technicians who have to produce all the labor to make that happen. They each work roughly 2080 hours per year. Times three = 6,240.

Now, calculate the $200,000 cost of operations divided by 6,240 = $32 per hour.

3. *The Complete Method.* Finally, there's the Complete Method for calculating labor costs. This involves looking up several numbers. You'll come up with a table that looks something like Table 8-1 below.

Labor Cost Calculator		Annual Cost
Base pay rate:	$20	
Hours worked per year:	2080	$ 41,600
Overtime hours worked per year: (Factor for overtime = e.g., 1.5 x base rate)	100	$ 3,000
Vacation and Holidays		$ ______
Employer's Contributions:		$ ______
FICA	6.20%	$ 2,765
Medicare	1.45%	$ 647
Federal Unemployment		$ 434
State Unemployment	1.00%	$ 446
Workmen's Compensation	7.00%	$ 3,122
Health Insurance		$ 2,000

Liability Insurance	$	300
Uniforms	$	______
Tools and Supplies	$	500
Training	$	500
Fair Share of Office Supplies	$	500
Computers and Equipment	$	500
Fair Share of Operating Expenses	$	500
Other	$	______
Total of All costs:	$	56,814
Calculate Total Annual Cost ÷ Total Hours worked =	$	26.06

Table 8-1

Note: Obviously these numbers are "made up" from some general calculations based on our industry.

When you choose to calculate the "Complete Method" you will need to go figure out your cost for rent, your cost for workers compensation insurance, etc.

Our experience has been that the more complete calculations of cost per employee or cost per hour of labor are lower numbers than the "quicker" methods. But, surprisingly, they're still in the ballpark.

The Final Labor Calculation has to do with your PSA (professional services administration) tool.

Here, you need to figure out how much of the total labor available to you is actually billed out to clients. Consider, time spent in:

- Meetings

- Trainings
- Rework
- Driving time
- Managing others
- Time spent in excess of estimates

Some companies get 1800 hours of billable labor from a technician in a year. Some get 1500. You need to know your number!

Note that lower level technicians will have more billable hours than higher level technicians. Our experience has been that lower level technicians top out around 75-80% billable.

Managers might be only 10-15% billable. High level technicians will tend to be in the 50-60% range.

Your PSA system should be able to generate a quick report that presents truly billable hours separated from all other hours.

(A moment of sad truth: Your PSA probably cannot spit out this number. After working for years with both ConnectWise and Autotask, we can tell you that neither of them can generate this number. But you can generate some reports and calculate this number yourself.)

What's the Final Calculation?

The final labor calculation is to multiply your labor estimate by the actual rate at which you can bill labor. Here's what we mean:

Let's say you have a technician who is 70% billable. And, let's say that his true cost per hour is about $30, based on the Super-Fast Method above.

You'll have to pay this technician for about 1.43 times as many hours as you can bill the client. In other words, ten hours of billable labor to the client will cost you 14.3 hours of labor to produce.

The true cost to provide an hour of labor from this technician is $30 x 1.43 = $42.90.

Now for the bad news.

Let's say you have a senior tech who is 50% billable and has a fully burdened labor rate of $50 per hour. For every hour you bill the client for this person, you have to pay her two hours.

The true cost to provide an hour of labor from this senior technician is $50 x 2.0 = $100.00.

You might have one round number for your team as a whole, knowing that some mix of hours will average a certain rate. Again, that knowledge comes from experience and documenting your experience.

Once again, let us remind you that your PSA system is an invaluable tool if you put the information into it in a manner that will allow you to get good reports out of it.

Check Point

Okay, here's where we are in the quoting process. You've defined the old network and the new network. From this you'll put together a list of things you need to sell to the client.

You also have some vague idea of what it will cost. The labor estimate is based primarily on your experience.

The keys to making sure you can complete this job profitably are to know as much as you can and to make sure the client knows that your labor estimates are estimates.

If you get in the middle of a project and the client comes up with a mission critical database that can't have downtime, you will have to make clear that all associated labor is outside the scope of the project.

Two Quotes

Now, let's look at the actual quote. We've generated some Excel spreadsheets so you can play with them. These two Quotes are included as the last two pages of this chapter. Review them as we go through them.

One of the quotes is marked "internal use." The other is the actual quote that's given to the client.

(Note: The Excel spreadsheets for these quotes are included in the downloadable content for this book. See p. 5 for more information on registering your book.)

As you can see, we divide the migration project into seven different categories and we create six or seven service requests in our ticketing system (two of the SRs are for training, so they can be combined). So, they will literally say "Stu Pidasso Migration Project" and the first one is training. The second one will say "Stu Pidasso Server Build." And so forth. The first item and the sixth item are "training."

The second item is an actual discovery process!

If this were not our client, and we are more or less coming in blind, we would probably put in three or four hours for the discovery process. But, if we're managing the client's computers, we simply need to ask "is there anything I don't know that's not in the PSA system?"

The third item is the server build. This includes everything from opening the box to installing the RAID controller, installing the operating system, and everything needed to prep the machine for the client's use. We have a standard build for a 2012 Server.

The fourth item is the network migration. This includes moving all the data, databases, etc. It normally includes getting old printers off the old server and onto a configuration we prefer. It includes reconfiguring the firewalls, network scanners, and anything else that touches the Ethernet.

If email is moving in-house, or to a spam filter, or moving out-of-house, it may also be included in this process. As a general rule, however, we like to separate such things so that they are two distinct projects and each has a definitive ending point.

The fifth item is the workstation migration. This includes all the work needed to get users and their data off the old system and onto the new. In this example, we have seven and a half hours for workstation migration. This is based on .75 hours per workstation. If they have a lot of old equipment it might go up.

And the seventh item? Did you spot it? It's the cleanup buffer. We add a buffer at the bottom just to make sure we're okay. Again, this is what we do internally. The client never sees this.

We all know how big projects are. Stuff happens. But we don't call it "stuff" and we don't call it "fallout." Think about it this way: You know something's going to go wrong. You don't know what it is right now, but something will go wrong. You also know that you'll find it, you'll fix it, and everything will be just fine. It might be your team's fault. Or it might be a piece of equipment that shipped with a broken part.

It doesn't matter.

Something will go wrong. You'll fix it.

You also know that there's a certain amount of overhead to running a project. In contracts with state agencies, they actually have a line of 10% or 15% for contract management. They recognize that that's the reality of running a project. We don't normally get to do that in the SMB space. So, you need some buffer. We start with 10% and tweak it up or down to come up with a nice round number.

In our example here, we have a total of 26 hours with a buffer of three. That's 29 total and we'll quote 30 to the client.

Approval Process

Remember the discussion of the sales roles? The sales person is putting together this estimate. And then, the sales person has to get approval of these estimates from the sales engineer.

You could, theoretically, set all these numbers to zero and make the sales engineer fill them in, but there's no real point to that. You should begin with estimates based on past jobs and let the sales engineer adjust the numbers based what the technical department knows about the clients, the operating system, the hardware, and the software.

Your process may vary. In fact, if you're not running our company, your process will vary. You might take the estimate to the service manager, the sales manager, or even the company president. Basically, you're hoping for an endorsement that it's a good estimate.

Note for growing companies: In the evolution of your company there will likely be a time when the one person who used to be the only person in the company is now primarily doing sales. The more you do sales, the less you deliver services. So, over time, your time estimates will be based less on experience and more on the advice of others.

Our internal rule is that we consider our estimate to be a cap on the number of hours we'll charge. We don't advertise that to the client, but it's what we do. We love flat-fee agreements for two simple reasons. First, we design them to be profitable under the worst case scenario. Second, if we find a better way of doing something, we become more profitable.

Note also that we put a major emphasis on staying inside the scope of the project. The result is that added work goes into separate service requests and is billed separately. All the labor inside the Network Migration Project is really just what you'll find on the checklist in this kit. That makes estimation a lot easier, and it makes sticking to the estimate a lot easier.

KPE nterprises
Sacramento, CA
KPEnterprises.com

This quote is valid for thirty days.

Quote

Client: **Internal Estimate** **May 2015**
Quote expires 6/30/2015

Project: **Server Migration -- 5 desktops**

Labor -- estimate

Qty			
1	Staff Training before installation: Introduction to the new system.	$ 150.00	$ 150.00
1	Estimated Labor for Discovery process: documenting current system, shares, users, etc. Document New system.	$ 150.00	$ 150.00
10	Estimated Labor for Server build: Set up server computer; configure Server 2012 Foundation; create users, shares; Install and configure backup software, Anti-Virus; configure DHCP, DNS, other services; create shares, etc. Coordinate MAS 90 setup and data transfer. Document and test.	$ 150.00	$ 1,500.00
5	Estimated Labor for Network Migration: Reconfigure firewalls, printers, routers, and additional equipment as needed. Restore data from disc/tape or otherwise transfer to new system. Note: Extremely large data transfers may add additional time.	$ 150.00	$ 750.00
7.5	Estimated Labor for Workstation Migration: Set up desktop PCs and connect to new server; configure; document everything. Note: this assumes there will only be minor cleanup needed on any machines, that machines are not "too" old, and that they are in reasonable condition.	$ 150.00	$ 1,125.00
1	Staff Training after installation: Becoming familiar with new features.	$ 150.00	$ 150.00
26	Sub total	Subtotal	$3,825.00
3.00	Add 10% buffer for hours for stuff they forgot to tell us.	$ 150.00	$ 450.00
29	Grand Total	Total	$4,275.00

8-1: The "Internal" Quote

KPEnterprises
Sacramento, CA
KPEnterprises.com

This quote is valid for thirty days.

Quote

Client: **Stu Pidasso, J.D.** — **May 2015**, Quote expires 6/30/2015

Project: **Server Migration**

Hardware and Software

Qty	Description		Amount
1	New Server. Includes: **Server - Proliant ML 350 Xeon 3.0 GHz, 16 GB RAM; 3x1 TB hard drives**; Drives configured in RAID 5 (2 TB total usable space); DVD+R/CDRW; Server 2012 Essentials		$ 5,799.99
		Sales Tax	$ 449.50
		Total:	$ 6,249.49

Labor

- Staff Training before installation: Introduction to the new system.
- Estimated Labor for Discovery process: documenting current system, shares, users, etc. Document New system.
- Estimated Labor for Server build: Set up server computer; configure Server 2012 Foundation; create users, shares; Install and configure backup software, Anti-Virus; configure DHCP, DNS, other services; create shares, etc. Coordinate MAS 90 setup and data transfer. Document and test.
- Estimated Labor for Network Migration: Reconfigure firewalls, printers, routers, and additional equipment as needed. Restore data from disc/tape or otherwise transfer to new system. Note: Extremely large data transfers may add additional time.
- Estimated Labor for Workstation Migration: Set up desktop PCs and connect to new server; configure; document everything. Note: this assumes there will only be minor cleanup needed on any machines, that machines are not "too" old, and that they are in reasonable condition.
- Staff Training after installation: Becoming familiar with new features.

30 hours $ 4,500.00

Note: the first three hours of labor will be credited if the server is on a SuperStar I.T. Service Agreement.

Thank You
for asking *KPEnterprises* to bid on this job!
Your Satisfaction is Guaranteed

Terms of Sale

All labor charges are non-refundable. * All hardware is sold with a manufacturer's warranty. If the manufacturer's warranty is less than 1 Year, KPEnterprises warranties the hardware from the end of the manufacturer's warranty to the end of 1 Year from the date of purchase. * All merchandise may be returned within 30 days unless otherwise specified. There is a 10% restocking fee on all items except special-order items. * There is a 20% restocking fee on special-order items. * Software licenses are not returnable/refundable. * There is a $25 charge on returned checks. * All unpaid sums that are not in dispute shall bear interest at the rate of 1.5% per month. Cost of collection, including reasonable attorney's fees, shall be borne by the client. * There is no additional charge for the use of a Visa or Mastercard to pay your invoice.

Approval: I Agree to these terms and approve the work described above:

Authorized signature ____________________ Date

8-2: The "External" Quote

Details

So, with good estimates and a 10% buffer, we come up with 29 hours for the project. Now, move over to the quote that's given to the client. We round 29 to 30 and we remove the item-specific time estimates. We keep the detail about what's included. In fact, we keep the basis for our estimates on the workstations. This does three things.

First, we've given ourselves a pool of thirty hours in which to profitably migrate the network. This gives us a lot of flexibility. So, for example, if the network migration portion only takes two hours but the workstations take an extra three hours, we'll be okay and don't need a change order from the client.

Second, we eliminate nitpicking. We avoid having the client ask why it takes so long to configure a firewall, or migrate a workstation. If one category is too big, a client might ask to just do that piece themselves. We all know that never works out, and rarely saves any money. It's better to simply lump it all together.

That big number of hours looks pretty reasonable when you've got those big blocks of text supporting it.

And third, we have left verbiage that gives you a basis for separating out work. For example (and this is the most common example), you find a workstation that's massively infested with viruses and spyware, loaded with data that should be on the server, AND it's five years old.

You're not going to migrate that machine for .75 hours labor. Since you've clearly laid out your criteria, you can go to the client and say that you'll create a service request to clean up that machine on a time and materials basis.

The second most common scenario is that you took the client's word about how many machines there are. And, on the day of migration, the number was way off. If you show up to migrate a 25 user network and face 35 machines, you can scurry over to the boss for either a change order or to create an SR for migrating those ten extra machines.

As you can see, we give the client two prices: One for the hardware and software, and one for the labor. Two dollar amounts. Even if they say, "Let's look at that hardware a little more closely," they're not likely to engage in a discussion about specific hardware requirements such as drive speed.

We give the client enough detail that they can make a decision, but not so much detail that we get embroiled in a lengthy discussion of the details. They need to trust that they are going to get a great server for a great price and that, three years from now, they are going to say you did the right thing for them. And they'll be ready to do it again!

Final Thoughts

This is obviously our process for quoting a network migration. Your process may be very different.

Our goals here are threefold:

1) We want to create a quoting process that is easily replicated from client to client (and from operating system to operating system, for that matter).

2) We want to make sure we're making money on every job. This involves a couple of things. First, we need to know our cost for delivery. Second, we need to have accurate estimates. The dissection of the project into smaller parts helps us achieve both of these.

3) We want to start building a Scope of Project so that we can agree with the client on what's inside the scope of the project. Everything else, of course, is outside the scope.

As we've presented it here, you should be able to start plugging in your numbers and building similar quotes for your own migration projects.

Many people take a very different approach to project quotes: They produce a 2030 page document that outlines why you chose this operating system, why you need RAID 5, etc. These tomes are not read by many clients and are primarily produced to create a thump factor when dropped on the client's desk.

Still, as one reviewer reminded us, analytical types (like CPAs and Engineers) really appreciate the details and want to understand as much as possible.

We prefer the shorter quote you see at the end of this chapter because the entire focus is on the project ahead. Everything fits on a page. Sign here, give us a check, and we're on our way.

As with all other business advice, integrate what works best for you and throw the rest away!

Final Steps / Approved Projects

Once you have a signed quote, your project begins.

In our case, we ask that all hardware and software be prepaid. Once we get a check for that, we create the service requests (six or seven) and order the equipment and software.

At this point, technicians can start working the discovery stage and gathering information.

It is also important that you begin the communication process with the client. When you've ordered the product and created the service requests, send the client an email that states what you've done and assures them that the project is moving along.

It is important to give the client an update at least once a week regarding the status of their project. Early on, they've given you a bunch of money and they don't see anything. You need to keep them confident that you're working on it.

If there are delays, make sure the client knows this. For example, if software hasn't arrived (especially the operating system), you can't really do much. Make sure the client knows that.

Next, we're going to talk about managing your project inside your PSA – Professional Services Automation – tool.

Your To-Do List for The Chapter

______ Download the Excel templates mentioned in this chapter and customize them for your business

______ Pick a "sample" project from this book – or a real one you have coming up for a client – and plug your numbers into the quoting template

______ Determine which estimate you will use for labor calculations. Enter a note about this into your quoting template

Notes

Chapter Nine: Running Your Project in a PSA

With luck, you've invested in some kind of ticket-tracking system or Professional Services Automation tool (PSA). You might also call it your service board. It could be something you built or something you bought.

This chapter covers how to manage the project inside the service board or PSA. We're going to use those terms interchangeably. You will just substitute whatever you're using.

We will not discuss all the other things you could be doing (should be doing) in your PSA, such as client management or contract management. We care about two things in this chapter:

- Tracking labor as accurately as possible
- Tracking progress for the project

Translating The Scope Into Tickets

The Scope of the Project IS the Project!

Remember the discussion from Chapter Three and other early chapters: Scope creep kills projects! Once you have an approved project, there is a natural progression from project definition to project completion.

In a perfect world, each of the line items in your Internal Quote will flow through to the External Quote you give to the client. And, in that perfect world, each of these line items will become a service ticket in your PSA or job board.

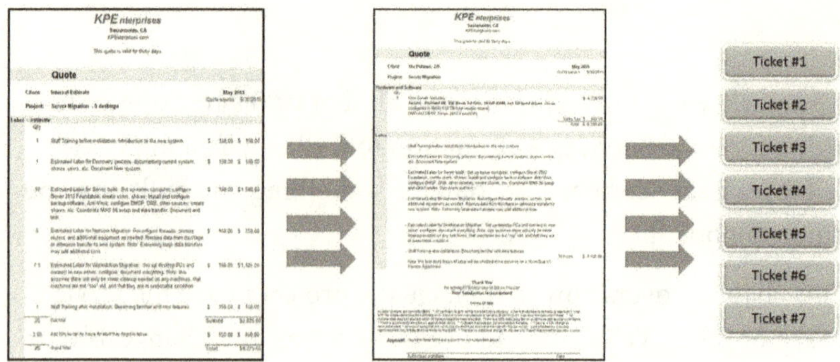

9-1: Flow from Internal Quote to External Quote to Tickets

So, let's say that you're migrating a server. You might have the following line items that become individual tickets:

1. Training (before and after)
2. Discovery
3. Server Build
4. Data Migration
5. Workstation Migration
6. Network Updates

Internally, you've already created time estimates for each of these items (see previous chapter). So, you can put time estimates into the service tickets. Two hours for training, ten hours for server build, etc. Resist the urge to re-estimate at this time. It's important that you put the same time estimates into the ticket as you have in your Internal Quote. That way you can track how accurate your estimates are!

Each ticket will have a Title or Short Description. It's a good idea to name your tickets so that they are all grouped together when you sort tickets by name. That way, this project won't get lost in your service board if you have hundreds of other tickets. For example:

- Migration Project 1 – Training
- Migration Project 2 – Discovery
- Migration Project 3 – Server Build
- etc.

At this point, you'll have a set of tickets that "are" this project inside your service board. The total hours allocated to these tickets represents the hours allocated to the project. And, of course, the total hours worked on these tickets represents the total hours worked on the project.

Tickets Outside the Scope

Karl likes to joke in project trainings that every technician should have two tattoos – one on each arm:

Inside the Scope

and

Outside the Scope

Every action you take for this client is either inside the scope of this project or outside the scope of the project. Every technician needs to understand that and understand the scope. When a user (or your principal contact) asks to do something related to the scope, they need to be very clear about what IS and what IS NOT inside the scope of the project.

In larger organizations, changes and additions are handled by a more formal "change order" process. But, in smaller businesses and smaller projects, we handle this much more easily. If something is outside the scope of the project, you simply create a new service request.

On the next page, we've reprinted the flow chart from Chapter Three.

Luckily, there are only a few kinds of issues that can come up during any SMB project. About 99% of all issues that arise fall into these three categories:

1. Fixed by the project. These are issues that are directly related to the project, but should be eliminated once the project is complete. For example, if someone has intermittent problems connecting to a server share, that better be fixed when the new server installation is complete.

2. So small you go ahead and include it. Sometimes a request will literally take a few minutes. For example, "Can you put a shortcut to that on my desktop?" You can go ahead and fold these niceties into the project.

But beware of "death by a thousand paper cuts." Some clients just have a way of adding tiny things again and again until they become a significant amount of labor. Your service manager or project

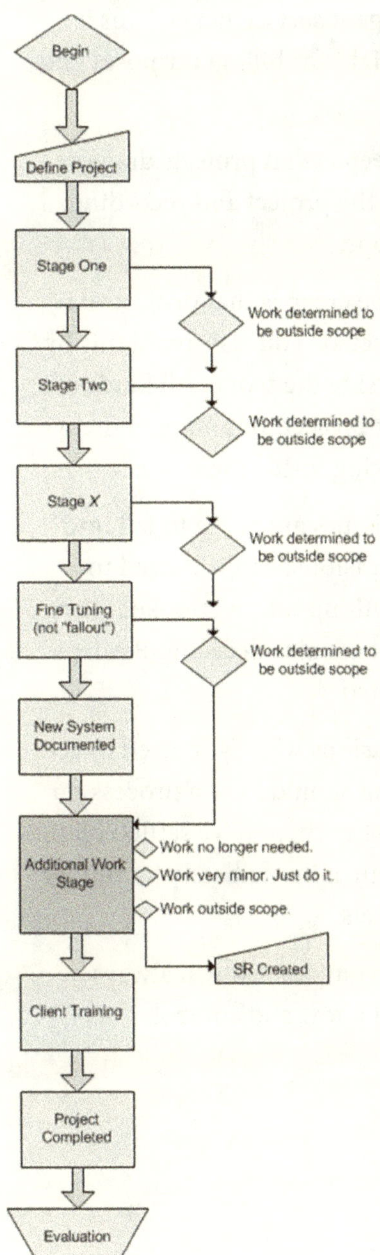

manager should be involved in all decisions to include these "little" things in the project.

Scope creep can creep up very quickly because it's a sneaky thing!

3. Clearly outside the Scope. All other requests are outside the scope of this project. The only option is to create a new service request (ticket).

At some point, your service manager will review these tickets with the client. Some will be urgent, some will be covered by a managed service agreement. Some will cost the client money. Some will rejected by the client as unnecessary.

All of them will be outside the scope of this project.

New Urgent Work

Let's say that one of these out-of-scope service requests must be executed in the middle of your project. It is critical that you keep the labor straight.

In other words, you need to stop entering time on the project tickets and start entering time on the new, urgent service ticket. This is critical because it keeps your time straight for billing purposes and for project management purposes.

One of the negative effects of scope creep is that projects drag on forever. You avoid that by suspending the project and recording time where it belongs – outside the scope.

The other major negative effect of scope creep is that projects become much more expensive than projected. You also avoid this by putting time on the tickets that are outside the scope. All kinds of things can come up and interrupt your project. But none of them should be able to affect your time working inside the scope.

Remember we said that 99% of "new" issues are going to fall into those three categories. On very rare occasions, you will need to modify the project itself. Perhaps something was overlooked. Perhaps the client changed his mind about a major decision. Perhaps the solution just doesn't work as expected.

Whatever the cause, there are rare occasions when you need to actually amend or modify the project. You should have a process for taking care of this. And, when you and the client have settled on the revised scope and price, you need to return immediately to putting all that information into a series of tickets.

No matter how far "off track" you get, your success will always be related to getting back on the track you know and control.

Put Time On The Right Ticket!

It takes some practice, but it's very important that everyone on your team enter time into the right ticket. Remember your quoting process: It became more accurate over time because you know how much time it actually took to perform each of these tasks.

If you go to get your car repaired, the dealership doesn't send someone out to poke around your car, scratch his head, and guess what it will cost to repair something. Instead, they have a list of tasks and the hours associated with them. In this way, they can combine any group of tasks into a repair job and know how long it should take.

They total up those hours and multiply by their hourly rate. That's your estimate! It's accurate and allows for them to make sure they're profitable.

This information is so standardized that you can estimate your own car repairs. For example, NAPA has an estimator you can use for free. Just put in your car make, model, year, and the job to be performed.

See

http://www.napaautocare.com/estimator.aspx

The point is, that you get accurate estimates because you know how long it took to complete these tasks before. And that means that it's important to track time correctly on every job.

Let's say that a technician is building the server and also in charge or reconfiguring the firewall. She cannot actually do two things at once, so she needs to be diligent about putting time on the right ticket. She might start the server install and log fifteen minutes to that ticket. Then she starts the firewall configuration as the server installation chugs away. After half an hour, she logs thirty minutes on the network configuration ticket and goes to check the server.

It takes teamwork and discipline to get in the habit of putting all the time for all the employees onto the right ticket. But, it's really in your long-term best interest.

Completing Tickets, Completing the Project

Once again let's just say: "All hail the mighty Scope of Project!"

As you complete and close each ticket, you complete each stage of the project. When all the tickets are closed, the project is complete.

The process we've followed from quoting the project to project approval, creating tickets, and working tickets has brought us to a logical conclusion of the project. Remember way back in Chapter Two we gave the PMBOK® definition of a project:

> ***"A temporary endeavor undertaken to create a unique product, service, or result."***

What makes a project go "bad" in the small business environment? They drag on forever. They cost more than expected. And for the consultant, both of those can mean that they are unprofitable. Yikes! (Actually, these concerns are the same in any environment.)

Using a systematic approach, such as the one we've described, you can complete all the goals of your project (the stages) on time and within budgeted hours. That means you're profitable.

Managing your service board is a completely separate discussion – maybe even a different book – but it's critical that you manage the board so that tickets don't languish without resolution. As you manage that board, you will also make sure that these project tickets don't sit there forever.

Few things make a client angrier than projects gone bad. And that's true even if it's the client's fault.

Few things make a client happier than a project that comes to a successful conclusion. So, make sure you close out those tickets and schedule a meeting to evaluate the project from the client's point of view.

In the next chapter, we'll wrap up the how-to discussion of projects. The final section of the book provides some sample projects you can use to begin creating templates for your own projects.

Your To-Do List for The Chapter

______ Using the sample project you picked for the "To-Do List" in the previous chapter, create service tickets for your project

______ Create some documentation for yourself. What are **your** steps for moving from "quote" to service tickets? Create a consistent, repeatable process.

Chapter Ten: Bring it All Together

Note: This is the concluding chapter on "how-to" theory. The final section consists of example projects.

When we start a big project, we always tell technicians "Something's going to go wrong. We don't know what it is. But we'll overcome it and be completely successful."

We know this to be true because it always has been. The same is true for you.

Stuff happens: you delete the router configuration; the client ordered the wrong drive cage; you discover a broken part at midnight and all the stores are closed.

But somehow you rebuild the configuration, mount the drive, and make the system work.

No matter what the project is, it won't go perfectly. But every time you go up against a problem, you learn a little bit. And when you do something without planning, you learn that the sequence of tasks matters. So you get better over time.

Project management – no matter how informal – is always going to be more successful than just jumping in and doing the work. We can't promise the twenty forms in the Project Binder are going to make every job perfect. But having a system for managing projects will always make them more successful than if you didn't have a system.

Here are a few ways that project management can improve your business in addition to simply improving your projects.

Better Sales

We actually use the project process to help with sales. We tell prospects that we're a step up from the competition because we have a *system*. We don't just go in and start poking around on their network.

That's also why we don't meet the prospect, have a five minute chat, and tell them what they need. Oh, no. We need to evaluate their systems, evaluate their goals, ask about their budget, and help them make the right decision.

It slows things down a bit, but businesses appreciate the fact that we have a process. It definitely differentiates us from the competition.

Did you ever witness a kid's birthday party organized by an amateur entertainer (clown, balloon artist, etc.)? They don't know quite what to do next. They have a skill and a bit of shtick, but are easily side-tracked when one kid says something that's not in their playbook.

But a professional entertainer will have a much better experience. They'll control the situation from A to Z, have alternative plans, and even have techniques to fall back on when things go wrong.

The difference for the people paying the bill is dramatic. Parents feel much better about giving their money to a professional than to an amateur. They get a sense that they can step back and let the pro do the work. They'll even pay a higher price because they know they'll get a better product.

Your business is the same way. An amateur will bumble around without a plan. A professional will have a plan, have some shtick, and be able to handle the entire process with a sense of confidence that is contagious. The people paying the bill will recognize this and pay a higher price for it.

Having a process can be a great sales tool. If you can walk in with a binder that's been filled out, with chicken scratches and a client sign-off, you can show a prospect that you know what you're doing. They can trust you to take on a project because you know how to get the job done.

The Royal We

Here's a great tip for dealing with clients and prospects: Learn the royal "we."

Rather than saying things like "I can figure out any router," or "I've worked with all these systems before," start your sentences with the word *we*. We can figure out any router. We have experience with all these systems.

And more importantly, make it sound as if you've handled so many projects like this that you have a system for everything.

- We like to control the DNS because . . .
- We run full backups every night because . . .
- We clean all the machines before joining them to the domain because . . .

You get the picture. When you talk, you need to come across as the pro who has a system for everything. And the truth is, you probably do. You have your "way" just as our companies have our "way" of configuring routers. You have a "way" for storing technical data, configuring backups, pushing Office to the desktops, configuring Blackberries, keeping track of documentation, etc.

You've got a system. You need to communicate in such a manner that the prospect sees you have a system. Maintain an air of confidence.

This all loops back to projects because projects are the way that you get things done. You don't just jump in and start making changes to the firewall. You don't click first and ask questions later. And you don't give a quote based on inadequate information.

Everyone's always in a hurry. They want the job to start today.

"But we like to make a plan. First, we do a discovery process. That way we understand your systems and minimize surprises. Second, we like to understand the bigger picture so we can make sure our recommendations fit in with your larger goals."

Minimizing Downtime

Projects also give you a huge advantage with regard to downtime.

If you make it your goal to minimize downtime in every project, you will find ways to be successful at that goal. And that, in turn, will become a very important calling card for you.

Many small businesses assume that there will be large chunks of downtime for system migrations, moving web sites, moving email servers, and similar jobs. Why is this? Well, first, that's been their experience in the past. And, second, that's the way most SMB consultants operate.

With minimal resources, downtime has been necessary for small businesses. Larger businesses can set up clustered servers and fail-over operations. These are generally too expensive for SMB clients.

But technology is setting us free. You don't have to schedule downtime if you can use some modern tools and services to avoid it. In addition, you need to dedicate yourself to figuring out how to accomplish this goal.

Minimizing downtime played a role in all three of the sample projects. When we start a new project, we always ask the question "Is there any reason we can't do this with zero downtime?"

Notice, we don't ask *can we* have zero downtime. Our assumption is that we can. So we turn it around and ask whether there's any reason we can't. This change in perspective has helped us win several jobs.

When we quote a network migration, we put in writing that we expect:

- Zero companywide downtime for email
- Zero companywide downtime for the server (databases, file access)
- Zero companywide downtime for Internet access
- and approximately one hour of downtime per desktop user

We accomplish these goals time after time. And we do it during business hours (8:00 AM to 5:00 PM). We don't work evenings and we don't work weekends.

Because we have a system. We have a *project*. There are adjustments for every client (meaning we have to do it a little differently each time), but we use our standard network migration project to achieve zero-downtime migrations every time.

Karl's technology company has not performed a network migration to a new server that involved company-wide downtime since 2002. And in all that time they've only performed one migration on a weekend – by client demand.

Any consultant can do this. You just have to have a system. And that means a project. Detail the steps you'll take, in the order you'll take them. Go slow. Be careful. Check your work. Document everything. Know what your skills are. Use the right tools.

As you can imagine, we use this as a selling point. There are no after-hours charges, so our projects are not overly-expensive. The company gets a promise of zero-company downtime, so they see the value in that. They don't always believe us, but it certainly separates us from the competition.

And after we migrate their entire office to a new server and new domain with zero downtime? They believe anything we tell them! They believe we can work miracles. Now that's a selling point.

Recommendations

Back in Chapter Three we discussed when to use a project binder. Now that you've seen the whole process, it is good to reconsider this. Over time you'll develop specific criteria for when you use a project binder.

Remember, there's a continuum between a simple *task* and a complicated *project*. All simple tasks belong on the service board, in service requests. All complicated projects belong in a project binder. Where you draw the line depends on how well you use your PSA system. Our PSA system is truly at the heart of our business. As a result, we can handle a lot of tasks as service requests and not formal projects.

In general, we use project binders when

- We have a major project (lots of money or very important to the client)
- The project is obviously complicated
- The project will take a long time
- The project has a number of obvious stages

At first, it takes some time to figure out whether something is a project. Once you have the Project Binder in your box of tools, you will begin to recognize projects that make sense for a binder.

To get started with your first project, look for a fairly comfortable project that you do at least two or three times a year. Next time, instead of jumping right in, take time to lay out the project. List the stages. Put them in order. Create a binder. Work the process.

You also want to look at the sample projects, such as the email move. Set a goal of zero downtime and figure out how you're going to do that. Once you customize it for your company and your procedures, you'll be able to create a "standard" project you can execute successfully and profitably every time.

Final Words

Standard operating procedures will set you free!

Having standardized forms and procedures can go a long way to improving your company's success and profitability. Projects are simply the next step in that process.

Good luck. If you have questions or feedback, please email us at *karlp@greatlittlebook.com* or *danajg007@gmail.com*.

Notes:

Section 5 - Sample Projects

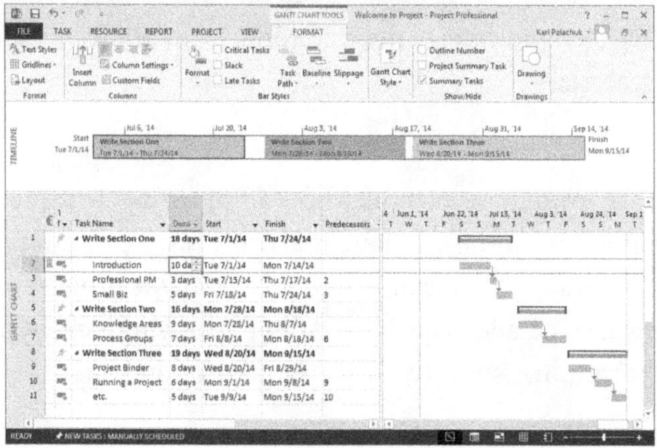

This section has three chapters. Each is a look at a specific SMB project in some detail, including filling out all the forms. The goal here is to give you some "real world" examples of how a project is run. We have selected three common, and reasonably simple projects. First, we present the forms from the previous chapter's example: Moving a web site from an in-house Small Business Server to a hosted web server. After that, we look at moving a client to a new ISP. Finally, we have a project to move from in-house email to hosted exchange mailboxes.

Note: The downloadable material for this book includes a bonus chapter with a sample project for bringing email in-house from a hosted environment. Since that's a much less common scenario these days, we decided not to include it in the book.

See the beginning of this book for information on how to register and receive the electronic forms and supplemental materials.

You can see from these examples how a project can become a repeatable process. And from there it is really just a procedure. Eventually, you could take one of these repeatable projects and simply create a detailed checklist. After all, moving a client from one ISP to another is mostly the same most of the time.

Why start over every time as if you've never tackled this job before? At the same time, it is important that you have a *process* and a *checklist*. Without those, you're likely to forget something. On every job, you'll forget one thing. But you'll forget a **different thing** every time you migrate a client!

With a process and a checklist, the only thing that can go wrong is failure to follow the process!

Note: To save some trees, I've trimmed out the tables of contents, the instructions for section five, and extra spaces at the ends of sections. In a binder, it is handy to have all this extra. But you don't need it here.

To take a look at these items, see Appendix A.

A page break for your binder is signified by a line like this:

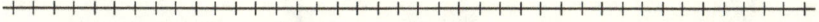

+-+

Chapter Eleven:
Sample Project - Moving Web Site from In-House To Hosted Server

Narrative:

This is the paperwork for the project we just discussed in the last chapter. Please note that these forms are meant to represent what the Project Binder looks like. Most of your detailed notes about what got done are in your ticket tracking system, whatever that may be.

We hope you get a sense that the timeline, as it gets filled out, will always give you the last action step and the next action step. These allow you to come up to speed very quickly.

And remember the flow: A few items came up along the way. These got added to the list, and separate service requests were created. Notice how that flows here.

KPEnterprises

Project Binder

Moving Web Site From In-House To Hosted Server

(Project Title)

For client Domain.com

Project Manager: Karl P.

Project Leader: Dana G.

Due Date: December 31, 2015

KPEnterprises Project Binder

Status

Current Status /

Last Action	Next Action Step	Date
Initiate project	Create Binder	12/26/2015
Created Binder	Get Due Date	12/27
Determine Due Date	Discovery	12/27
Did discovery: SBS !!!		
Fleshed out Stages	Get Approval	12/27/2015
Got approval		12/28
Stage One, Two Three	Talk to users	12/28
Stage 4-5-6-7-8-9	Wait for WWW	12/31
Done until www		
Repointed DNS		
Review project with KP	Done !!!	1/1/16

Table of Contents

(removed)

Project Description

Overview

Describe the primary goals of this project and how they will be achieved.

Move web server from in-house to hosted at Cousin Larry's Pretty Good Hosting

Already experienced downtime. So we need to make sure there's no unscheduled

Downtime. Need to make sure there's zero interruption in email.

2.0 Define Goals for the Project

Goals should be visible, attainable actions or results. You should be able to point to something and say: "This represents the completion of the goal." So, in addition to stating each goal, also state how you will verify completion of the goal.

Move web server from in-house to hosted at Cousin Larry's Pretty Good Hosting

-- web site will be up at new location

All functions of existing ~~Exchange~~ SBS server will continue uninterrupted

If new https certificate is needed, it will be generated

Users may need to be reconfigured

-- success = all users up on new connections to exchange

3.0 Define where the project (network, etc.) is today. What is the status quo?

Document current status. This may include descriptions, summaries, network maps, etc.

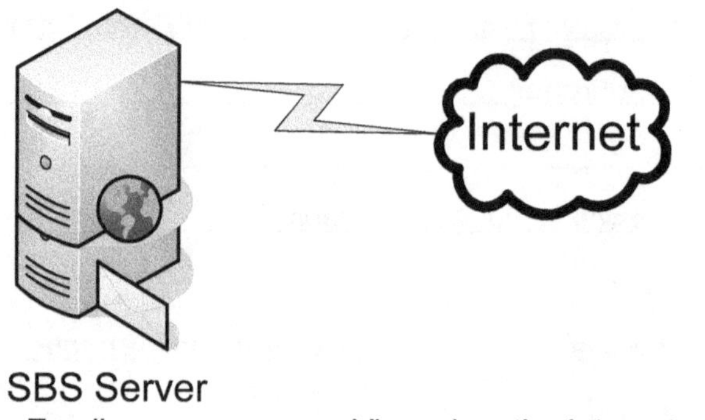

SBS Server

- Email	Viewed on the Internet as
- Web Site	www.domain.com
- RWW	www.domain.com/remote
- OWA	www.domain.com/exchange
- SharePoint	www.domain.com:444

IP = 123.234.221.5

NATs to 192.168.22.55 / 0

Firewall open on 80, 443, 444, 110, 25, 4425

4. 0 Define where you want to be.

Document what the system will look like when you are finished. That is, after you have met all goals, what will it look like?

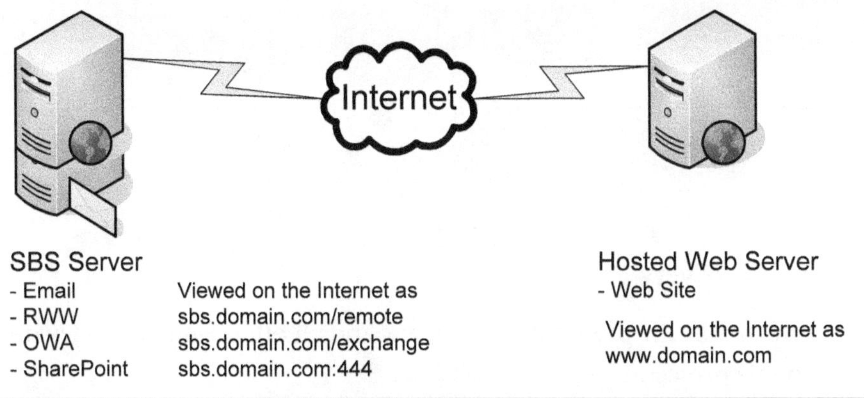

The following is unchanged.

IP = 123.234.221.5

NATs to 192.168.22.55 / 0

Firewall open on 80, 443, 444, 110, 25, 4425

5. 0 Define all stages and steps needed to meet all objects and goals.

(Removed)

5.1 Discovery

Document the pre-existing network and all other elements relevant to this project.

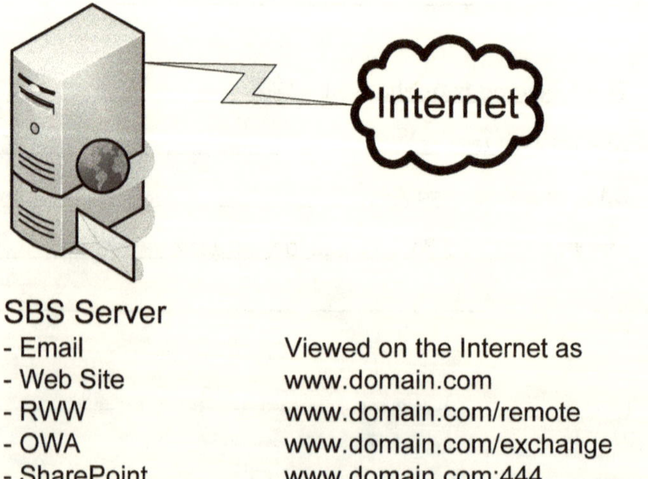

No VPN. Three remote users.

Previous section (4) updated to reflect SBS rather than Server 2003

Will use SBS.domain.com for /remote and /exchange as well as certificate.

Will train Victoria and she will train others on switch to sbs.domain.com from www.domain.com.

5.2. Define Training Needed by Client or Technicians

No training needed by techs.

Three users – Mo, Moe, and Joe – will need a little help with new certificate.

5.3. Timeline

Action, Stage, or Deliverable
Due Date

Begin project

12/27/2015

DNS changes / MX / letter to Larry

Find outlook and SharePoint users who need help

Update documentation

Change secure certificate in SBS

Test test test

Help users from above

Note: All in-house changes complete

12/31/2015

Re-point DNS to new WWW server

Depends on Larry

Document Everything When completed

5.4. Stage One

Service Request # 8596

5.4.1. Actions to be taken:

Create DNS entry for sbs.domain.com

Point to existing server (see documentation at front of binder)

Test

5.4.2. Notes:

-- no problems.

5.4.3. Additional Work / Issues.

Enter here and then copy to Final Stage

-- none --

5.4. Stage Two

Service Request #

(Repeat this page as needed until all necessary stages are defined.)

5.4.1. Actions to be taken:

Change mx records from www.domain.com to sbs.domain.com

Verify through DNSSTUFF.COM

Document in Network Documentation Binder

5.4.2. Notes:

-- No problems

5.4.3. Additional Work / Issues.

Enter here and then copy to Final Stage

-- none --

+++

5.4. Stage Three

Service Request #

(Repeat this page as needed until all necessary stages are defined.)

5.4.1. Actions to be taken:

Inform Cousin Larry's Pretty Good Hosting Service, in writing, that they need to stop any DNS hosting that they may have created for this client. Anything they

have could make a mess, even if it's only for their own customers

5.4.2. Notes:

Done. Received confirmation email. Both messages inserted in binder.

5.4.3. Additional Work / Issues.

Enter here and then copy to Final Stage

-- Just keep an eye on it. No SRs needed.

5.4. Stage Four

Service Request # ________

(Repeat this page as needed until all necessary stages are defined.)

5.4.1. Actions to be taken:

Identify all users who access Outlook using RPC over https. They will need the

new certificate and a configuration change. Inform them that a change is

coming. Identify users who will need help.

5.4.2. Notes:

Mo, Moe, and Joe need this. Contact information in PSA system.

See also SharePoint (next stage)

5.4.3. Additional Work / Issues.

Enter here and then copy to Final Stage

-- Mo and Moe have issues with Outlook staying connected.

Create an SR.

+++++++++++++++++++++++++++++++++++++++

5.4. Stage Five

Service Request #

(Repeat this page as needed until all necessary stages are defined.)

5.4.1. Actions to be taken:

Identify all user who access SharePoint. They will need to

know to install the

new certificate. Inform them that a change is coming.

Identify users who will

need help.

5.4.2. Notes:

Victoria says almost everyone uses SharePoint. We need

to help Mo, Moe, and Joe.

She'll train everyone else.

5.4.3. Additional Work / Issues.

Enter here and then copy to Final Stage

-- None --

5.4. Stage Six

Service Request # ________

(Repeat this page as needed until all necessary stages are defined.)

5.4.1. Actions to be taken:

Update any client how-to documentation re: Outlook, SharePoint, OWA, and RWW.

If it doesn't exist, create it.

5.4.2. Notes:

No documentation existed. Revised our standard handouts with specifics about

sbs.domain.com. Gave these to Victoria to copy and distribute.

5.4.3. Additional Work / Issues.

Enter here and then copy to Final Stage

-- none --

5.4. Stage __Seven__________

Service Request # __________

(Repeat this page as needed until all necessary stages are defined.)

5.4.1. Actions to be taken:

Change secure certificate so that it reflects the FQDN sbs.domain.com.

5.4.2. Notes:

Ran the Wizard. Tested.

5.4.3. Additional Work / Issues.

Enter here and then copy to Final Stage

-- none --

5.4. Stage Eight

Service Request # ________

(Repeat this page as needed until all necessary stages are defined.)

5.4.1. Actions to be taken:

Test.

Test that email flows in and out with new mx records. KP

Test RWW with new certificate. KP

Test OWA with new certificate. KP

Test SharePoint with new certificate. KP

Configure an Outlook client and test Outlook connection with RPC over HTTPS. KP

5.4.2. Notes:

All Good! KP

5.4.3. Additional Work / Issues.

Enter here and then copy to Final Stage

-- None --

+++

5.4. Stage Nine

Service Request #

(Repeat this page as needed until all necessary stages are defined.)

5.4.1. Actions to be taken:

For all clients who need assistance with the new configurations, help them get new certificate installed and working.

5.4.2. Notes:

Mo, Moe, and Joe all brought in laptops. Configured them all. Tested. Rebooted and had them demonstrate success.

5.4.3. Additional Work / Issues.

Enter here and then copy to Final Stage

Joe would like to upload a folder full of files to SharePoint. Not sure how. NOTE: This is outside the scope of the project. Create a service request for this.

5.4. Stage Ten

Service Request # ______

(Repeat this page as needed until all necessary stages are defined.)

5.4.1. Actions to be taken:

After new web site is up, change DNS to point www to new IP.

5.4.2. Notes:

Got email from Larry. Changed IP. Will verify in an hour.

KP

-- checked out fine. Had client test. All Good. KP

5.4.3. Additional Work / Issues.

Enter here and then copy to Final Stage

-- None --

5.4. Stage Eleven

Service Request #

(Repeat this page as needed until all necessary stages are defined.)

5.4.1. Actions to be taken:

Document everything in the Network Documentation Binder.

5.4.2. Notes:

Documentation completed. Network documentation binder up to date. KP

5.4.3. Additional Work / Issues.

Enter here and then copy to Final Stage

-- Done!!!

5.5. Fine-Tuning to make sure all stages are complete and all goals accomplished

Check with client, end-users, vendors, etc. Verify that all individual stages of work are complete and that they all work well together.

5.5.1. Notes:

-- Everything looks good. No additional work expected.

KP

5.5.2. Additional work or fine-tuning required.
If work is minor, define and perform it here. If it is more involved, copy to the Final Stage.

-- From Stage 9: Joe would like to upload a folder full of files to SharePoint. Create an SR.

Not Sure how.

-- from stage four: Mo & Moe have issues with Outlook staying connected. Create an SR.

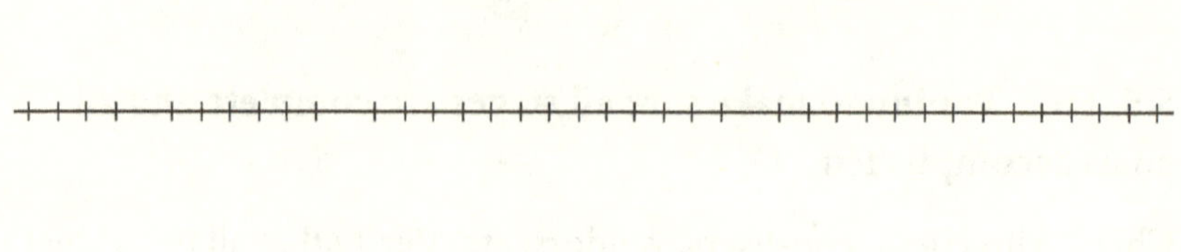

5.6. Documentation

All work should be documented as we go through the stages.

At this point, we need to verify that all documentation is complete, both in the Network Documentation Binder and in PSA system.

5.6.1. Notes:

Done. KP KP

5.7. Additional Work Required

Use this page to define all "additional" work that comes up during the project.

Determine whether the work is **inside** or **outside** the scope of the original project.

Determine whether work will be completed now (minor work only) or placed into a separate Service Request or Project.

5.7.1. Additional Work

Work Description	**Disposition / SR**
Joe would like to upload a folder full of files to SharePoint	SR 5995
Mo having issues with Outlook staying Connected	SR 5996
Moe having issues with Outlook staying Connected	SR 5997

5.7.1. Notes:

-- None.

5.8. Client Training

Complete Client Training as defined in 5.2. above.

Notes:

Training completed 12/31. Victoria will handle future training. She'll call if she needs us.

5.9. Project Completed and Signed Off

Verify with client that project is completed, all Service Requests are closed (completed).

Notes:

Talked to client. They're happy. Web is up. Email's flowing. SharePoint's working.

OK to close project. KP

6. 0 Final check-off for project plan

Meet with the decision-making stakeholders and verify that the plan is correct, that all objectives will be met, and agree on what constitutes the scope of the project.

After this step is complete, you may begin implementing the plan.

Met with Victoria on 12/27. Hours okay. Timing good. Set due dates.

OK to proceed. KP

7.0. Project Evaluation

Project manager and all technicians go through the project and determine what went well, what needs improvement, etc.

Notes:

Very good. Probably do trainings as a group in the future. Victoria had more trouble than she expected. After doing three individual trainings, it was clear that others needed help as well.

Chapter Twelve:
Sample Project - Switch to a New ISP

Narrative

In this example, we're moving a client to a new ISP. Our goal is to minimize downtime. The client's firewall is old, so they're willing to put in a new one. That will make the transition easier.

If you take the time to think about this, you can see where a little extra money can be squeezed out of this project. Specifically, we have a technician who is coming up to speed on firewalls. So, once you have the IPs from the ISP, you can let the tech start playing with the device. Check his work. Double check it. All at your office and not interrupting the client.

Show the tech how to back up the configuration, then start over from scratch and configure it exactly as it should be at the client's office. Save the configuration again.

So, you get a little training time with real world equipment and the firewall configuration is done at your site and not at the client's. If you are charging a flat fee, or this is covered under a service agreement, then doing the work at your shop is generally more profitable.

Another item of interest is the timeframe. In this case, you were looking for a tighter time frame (two weeks), but the old ISP needs

30 days' notice. That actually works fine. So you revise the time line so that notice goes out at the end of one month. You keep your two week window. And you still have the old ISP connection in place for an additional two weeks in case something goes wrong with the new setup.

Sweet!

KPEnterprises

Project Binder

Move TomCat Productions to New ISP

(Project Title)

Project Manager: Bob

Project Leader: Pat

Due Date: 4/1/2015

KPEnterprises Project Binder

Status

Current Status / Last Action	Next Action Step	Date
Define Project / Build Binder		2/25/08
Discovery done		
Wrote notice to ISP	Move DNS	2/25
Verified DNS		3/1
Got IPs. Coordinated reverse lookup	Install	3/6
Present for Install		
Configured firewall	Move DNS	3/12
Changed DNS. Tested.		
Memo to staff. SR 2002	Remove old Router at end/mo	3/14
Removed old equp.	DONE	3/28/2015

Table of Contents

(removed)

1.0 Project Description

Overview

Describe the primary goals of this project and how they will be achieved.

Client moving to new ISP. Need to minimize downtime.

willing to buy new firewall.

Major Goal: Zero Business Down.

2.0 Define Goals for the Project

Goals should be visible, attainable actions or results. You should be able to point to something and say: "This represents the completion of the goal." So, in addition to stating each goal, also state how you will verify completion of the goal.

Client has web server at 217.232.233.234 and email server at 217.233.234.235. RDP to citrix server at 217.232.233.236.

Looking for Zero Downtime.

Move all functions to new IP addresses.

3.0 Define where the project (network, etc.) is today. What is the status quo?

Document current status. This may include descriptions, summaries, network maps, etc.

Net map:

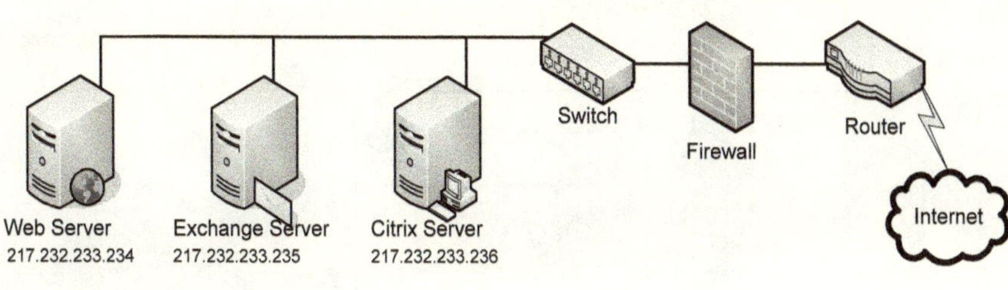

4. 0 Define where you want to be.

Document what the system will look like when you are finished. That is, after you have met all goals, what will it look like?

Map One: Interim step:

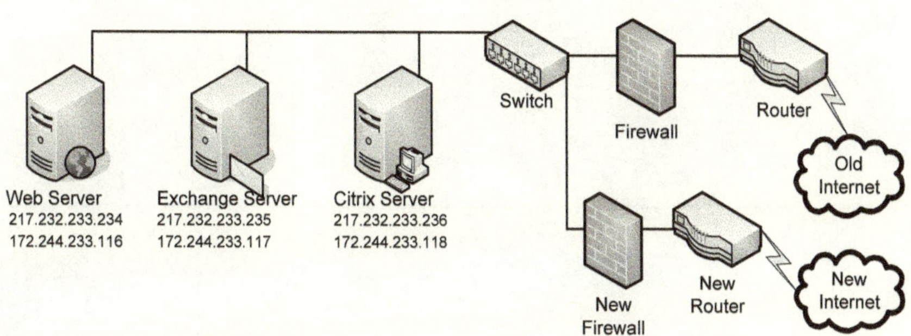

Map Two: Final setup:

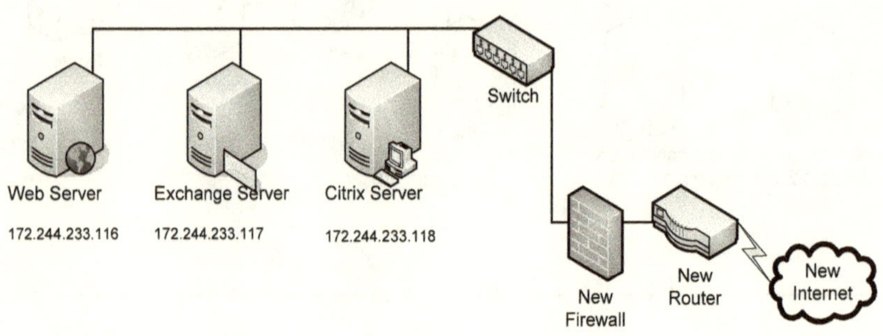

5. 0 Define all stages and steps needed to meet all objects and goals.

(Removed)

5.1 Discovery

Document the pre-existing network and all other elements relevant to this project.

See section 3.0

Network Documentation Binder is up to date.

Summary of Server/NAT:

217.232.233.234	Web Server	Ports	80, 443
217.232.233.235	Exchange Server		110, 25, 80, 443
217.232.233.236	Citrix Server / TS		80, 443, 1494, 2512, 2513, 22, 1723, 3389

5.2. Define Training Needed by Client or Technicians

Clients: Won't see any visible changes. No training required.

Techs: Alan coming up to speed on firewall config. Will work with Pat on that stage.

5.3. Timeline

Action, Stage, or Deliverable	Due Date
Give Notice to old ISP (30 days)	2/25/2015
Verify access to DNS. Move or get if needed	March 1
Get IPs from new ISP	Mar. 7
Coordinate Reverse DNS with new ISP (exchange)	Mar. 14
Verify service installed. Test	
Configure new firewall and server GW. Test	Mar 18
Change DNS. Test. On a Friday at 4 PM!	3/21
Memo to Clients re: access via IP vs. Name	3/21
Remove old firewall and Router	4/1/08

+++

5.4. Stage One **Service Request # 1995**

5.4.1. Actions to be taken:

Give Notice to old ISP.

5.4.2. Notes:

They require a 30 day notice.

Therefore, must be done by March 31st.

-- Done 3/25/2015 PL

5.4.3. Additional Work / Issues.

Enter here and then copy to Final Stage

+++

5.4. Stage __Two__

Service Request # __1996__

(Repeat this page as needed until all necessary stages are defined.)

5.4.1. Actions to be taken:

Verify access to DNS

-- Move to our control.

5.4.2. Notes:

First, made sure DNS was not at old ISP. Turns out, it is at Network Solutions. No problem.

Got access via client username/pw. Added ourselves as technical contact.

Reviewed DNS. All clean except references to old server IP no longer used.

Removed this. PL

5.4.3. Additional Work / Issues.

Enter here and then copy to Final Stage

Domain expires in four months. Create SR to get client to renew.

+++

5.4. Stage Three

Service Request # 1997

(Repeat this page as needed until all necessary stages are defined.)

5.4.1. Actions to be taken:

Get IPs from new ISP

5.4.2. Notes:

Done Mar 6 Entered into PSA PL

172.244.233.114 - 120

116 = server 1

117 = server 2

118 = server 3

119 = firewall

5.4.3. Additional Work / Issues.

Enter here and then copy to Final Stage

5.4. Stage Four

Service Request # 1998

(Repeat this page as needed until all necessary stages are defined.)

5.4.1. Actions to be taken:

Coordinate RDNS (reverse DNS) for Exchange (172.244.233.117)

Set up Service Record in DNS for Exchange (SRV record)

5.4.2. Notes:

See Service Ticket.

Call ISP. The person you get won't understand what you want. Keep asking

To talk to the person who actually handles DNS changes. That person will Understand. Must be complete by 3/14

5.4.3. Additional Work / Issues.

Enter here and then copy to Final Stage

5.4. Stage Five

Service Request # 1999

(Repeat this page as needed until all necessary stages are defined.)

5.4.1. Actions to be taken:

Verify new service installed.

-- If possible, be there on install date to plug in PC, verify access, and verify Correct IPs, verify that they're static, etc. Don't forget RDNS lookup.

5.4.2. Notes:

Get configuration information from ISP rep.

Create router configuration page. Label bottom of router.

Plug in laptop and access internet. Run DNS Stuff reports, including RDNS.

Verify that paperwork says these IPs are static.

5.4.3. Additional Work / Issues.

Enter here and then copy to Final Stage

5.4. Stage Six

Service Request # 2000

(Repeat this page as needed until all necessary stages are defined.)

5.4.1. Actions to be taken:

Configure new firewall. See config outlined above.

Verify access through all necessary ports to all IPs / services.

Label firewall. Fill out Configuration sheet for doc binder

Label bottom of firewall.

Backup firewall configuration to c:\!Tech on primary server.

5.4.2. Notes:

5.4.3. Additional Work / Issues.

Enter here and then copy to Final Stage

5.4. Stage Seven

Service Request # 2001

(Repeat this page as needed until all necessary stages are defined.)

5.4.1. Actions to be taken:

Change DNS on domain and gateways on servers

Do this at 4 PM on a Friday. Will take some time to propagate

5.4.2. Notes:

Inbound to either old or new IPs (and names) should be successful at all times.

Only problem is outbound response on web server. Since it will use first gateway,

There could be an inbound request from one set of IPs that is answered on

Another. This switchover problem will not last long.

Reminder: ALL DNS issues are typos!!! Double check everything. Go slow.

5.4.3. Additional Work / Issues.

Enter here and then copy to Final Stage

5.4. Stage Eight

Service Request # 2002

(Repeat this page as needed until all necessary stages are defined.)

5.4.1. Actions to be taken:

Write memo to client employees re: IP changes.

-- We are making a change that should be invisible.

-- If you access any resources by IP address instead of name, contact us

Immediately. If something stops working, open a ticket and we'll troubleshoot.

-- Expect possible troubles at the end of day on Friday. Everything should be Perfect by 8:00 AM on Monday.

5.4.2. Notes:

5.4.3. Additional Work / Issues.

Enter here and then copy to Final Stage

5.4. Stage Nine

Service Request # 2003

(Repeat this page as needed until all necessary stages are defined.)

5.4.1. Actions to be taken:

Remove old Firewall

Remove old Router.

5.4.2. Notes:

Leave up in case of emergency switch-back until end of month.

Document all work. Verify that all configurations are up to spec.

5.4.3. Additional Work / Issues.

Enter here and then copy to Final Stage

5.5. Fine-Tuning to make sure all stages are complete and all goals accomplished

Check with client, end-users, vendors, etc. Verify that all individual stages of work are complete and that they all work well together.

5.5.1. Notes:

5.5.2. Additional work or fine-tuning required.
If work is minor, define and perform it here. If it is more involved, copy to the Final Stage.

Domain expires soon. Create SR to get client to renew

5.6. Documentation

All work should be documented as we go through the stages.

At this point, we need to verify that all documentation is complete, both in the Network Documentation Binder and in PSA system.

5.6.1. Notes:

5.7. Additional Work Required

Use this page to define all "additional" work that comes up during the project.

Determine whether the work is **inside** or **outside** the scope of the original project.

Determine whether work will be completed now (minor work only) or placed into a separate Service Request or Project.

5.7.1. Additional Work

Work Description	Disposition / SR
Domain expires soon. Create SR to get client to renew.	2102

5.7.1. Notes:

5.8. Client Training

Complete Client Training as defined in 5.2. above.

Notes:

5.9. Project Completed and Signed Off

Verify with client that project is completed, all Service Requests are closed (completed).

Notes:

Done 4/3/2015

6. 0 Final check-off for project plan

Meet with the decision-making stakeholders and verify that the plan is correct, that all objectives will be met, and agree on what constitutes the scope of the project.

After this step is complete, you may begin implementing the plan.

Meet with client and get approval on all these steps.

2/25: Client requested additional step. Write memo to employees about

Changes we're making and their impact on users.

This was added on day of final switchover.

7.0. Project Evaluation

Project manager and all technicians go through the project and determine what went well, what needs improvement, etc.

Notes:

Generally very smooth. This is a good ISP to work with.
They're not greedy about DNS. Quick response on reverse lookup.

Alan did great on the firewall.

Chapter Thirteen:
Sample Project – Move Email from In-House to Hosted Exchange

Narrative:

Email projects are always interesting because they're always **zero downtime** for the client's email system. Or they should be. Here's the background:

Client has in-house Exchange email (could be SBS or stand-alone Exchange server). They are moving to hosted Exchange mailboxes. The specific example discussed here is Rackspace, but it could be just about any hosted Exchange service.

As part of this project, we're going to make this simple and secure by adding Hosted Spam Filter to the mix. If you're not familiar with the advantages of a hosted spam filter, this project should shed some light on that.

At some point, all email will go to the hosted spam filter. If no server is ready, the filter will just hold the email. Then we bring the new server online and re-point the email forwarding. Instantly, the new server is live.

Poof. Zero downtime.

KPEnterprises

Project Binder

Move BobCat, Inc. Email to Hosted Exchange

(Project Title)

Project Manager: **Neil**

Project Leader: **Robin**

Due Date: **7/15/2015**

KPEnterprises Project Binder

Status

Current Status / Last Action	Next Action Step	Date
Initiate Project		
Build Binder	Discovery	6/30/2015
Discovery		7/1
Set up Hosted Spam Filter to old site		7/2
Set up client and mailboxes at Rackspace		7/5
Cleanup email boxes on Exchange		7/6
Cleanup email boxes on Exchange		7/7
Export mailboxes	Stop Exchange	7/8
Move individual desktops to new server		7/9
		Thru 7/10
Point spam filter to new server		7/10
Verify email flows in and out		7/10
Train clients		July 11
Job Done!!!		July 11

Table of Contents

(removed)

Project Description

Overview

Describe the primary goals of this project and how they will be achieved.

Move all Email from in-house Exchange to hosted Exchange Mailboxes.

Put client on Hosted Spam Filter

2.0 Define Goals for the Project

Goals should be visible, attainable actions or results. You should be able to point to something and say: "This represents the completion of the goal." So, in addition to stating each goal, also state how you will verify completion of the goal.

Client has SBS2003. Exchange is in-house.

We are going to move all email to external hosted

Exchange mailboxes

All email will be consolidated and boxes "shrunk down"

In this process.

Integrate Hosted Spam Filter for spam filtering (in-coming)

Zero downtime for transition.

3.0 Define where the project (network, etc.) is today. What is the status quo?

Document current status. This may include descriptions, summaries, network maps, etc.

This is post-discovery:

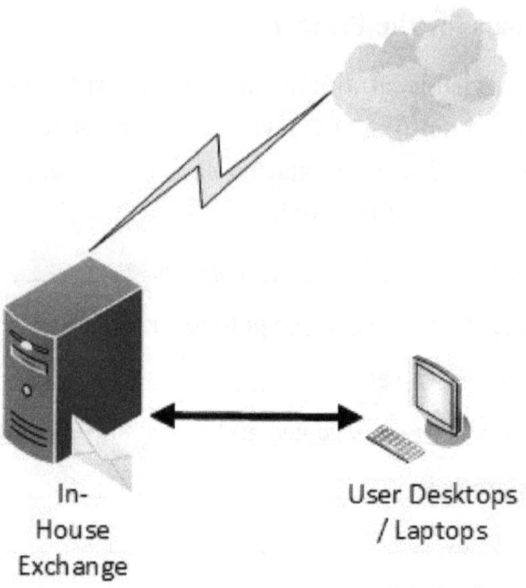

Email Environment "Before" Project

All users connect to in-house Exchange server for email,
Calendars, etc.
Remote users simply connect to RPC over HTTPS to access
Exchange server

There is no spam filter

4. 0 Define where you want to be.

Document what the system will look like when you are finished. That is, after you have met all goals, what will it look like?

<u>Interim Step – add Hosted Spam Filter</u>

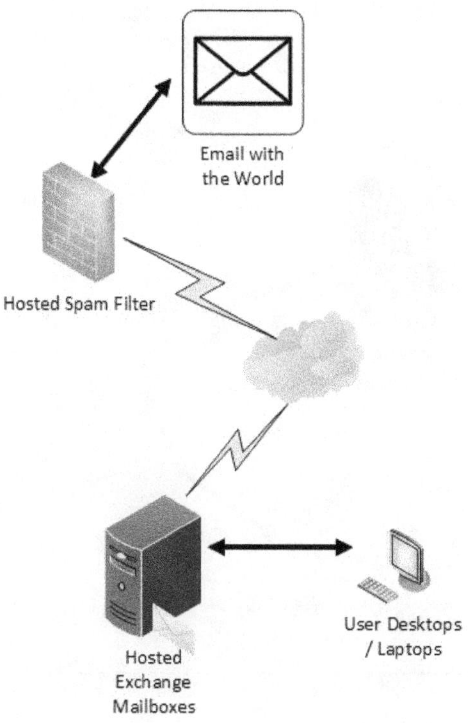

Final Setup:

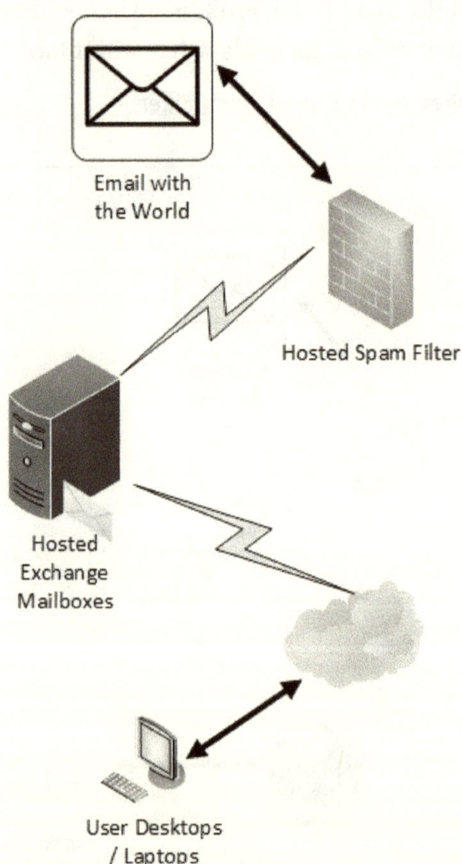

All Email to Hosted Spam Filter.

From Hosted Spam Filter to hosted Exchange mailboxes

All users connect to Exchange with Outlook.

+++++++++++++++++++++++++++++++++++++++

5. 0 Define all stages and steps needed to meet all objects and goals.

(Removed)

+++++++++++++++++++++++++++++++++++++++

5.1 Discovery

Document the pre-existing network and all other elements relevant to this project.

See Section 3.0.

+++++++++++++++++++++++++++++++++++++++

5.2. Define Training Needed by Client or Technicians

Client: Train group on OWA.

-- Memo with how-to information

Tech: No additional training required.

5.3. Timeline

Action, Stage, or Deliverable	Due Date
1. Set up Hosted Spam Filtering	7/2
2. Set up Site on Rackspace	7/5
3. Set up mailboxes on Rackspace Including users, aliases, contacts, and public folders	7/5
4. Clean up mailboxes on Exchange Server	7/6
5. Use Exmerge to export PSTs	7/7
6. For each user, create new Outlook profile On hosted exchange server	7/7
7. For each user, import PST and NK2	7/7
8. Verify all email, calendars, etc. are correct	7/7

9. Set Rackspace as default profile 7/7

10. Point spam filter to Rackspace 7/7

++

5.4. Stage One

Service Request # 1572

5.4.1. Actions to be taken:

Control DNS

Registrar is SRS Plus (www.srsplus.com). Credentials are in PSA.

Verify that DNS is at Dream Host (dreamhost.com)

Verify we can access DNS. Record all credentials in PSA

5.4.2. Notes:

Don't let this slow us down.

5.4.3. Additional Work / Issues.

Enter here and then copy to Final Stage

5.4. Stage ___Two________

Service Request # ___1573__

(Repeat this page as needed until all necessary stages are defined.)

5.4.1. Actions to be taken:

Set up Hosted Spam Filter and point to existing in-house Exchange Server

5.4.2. Notes:

5.4.3. Additional Work / Issues.

Enter here and then copy to Final Stage

+++

5.4. Stage ___Three___

Service Request # ___1573___

(Repeat this page as needed until all necessary stages are defined.)

5.4.1. Actions to be taken:

Set up email at Hosted Exchange Provider

Create accurate list of existing users and their emails.

Create all users and related aliases at hosting provider

Including users, aliases, contacts, and public folders

5.4.2. Notes:

5.4.3. Additional Work / Issues.

Enter here and then copy to Final Stage

One new user will be set up on Monday.

Outside the scope of this project. Create SR.

5.4. Stage Four

Service Request # 1574

(Repeat this page as needed until all necessary stages are defined.)

5.4.1. Actions to be taken:

Clean up existing mailboxes.

See "Mailbox Cleanup Procedure" under SOP directory

Archive old, deleted, emails. Place archives on the server

In the "Migration" directory.

Verify that all mailboxes are 2GB or less. Ideally, 100 MB

Or less. Document exceptions.

5.4.2. Notes:

-- Bob and Tom have extremely large mailboxes. We Cannot use Exmerge for these. Watch them carefully!

5.4.3. Additional Work / Issues.

Enter here and then copy to Final Stage

5.4. Stage Five

Service Request # 1575

(Repeat this page as needed until all necessary stages are defined.)

5.4.1. Actions to be taken:

Change DNS. Point to Hosted Spam Filter

-- Leave the dual connection in place for a few days.

5.4.2. Notes:

5.4.3. Additional Work / Issues.

Enter here and then copy to Final Stage

5.4. Stage Six

Service Request # 1576

(Repeat this page as needed until all necessary stages are defined.)

5.4.1. Actions to be taken:

Export email to PSTs.

For Bob and Tom, export from Outlook direct to PST.

- Verify that these PSTs are good – open with another Profile.
- Store PSTs on the server in the Migration directory

Export all other mailboxes via Exmerge

- Store PSTs on the server in the Migration directory

5.4.2. Notes:

5.4.3. Additional Work / Issues.

Enter here and then copy to Final Stage

5.4. Stage ___Seven___

Service Request # ___1577___

(Repeat this page as needed until all necessary stages are defined.)

5.4.1. Actions to be taken:

Import PSTs to email system at Rackspace

Follow the checklist. For each user . . .

- Create new Outlook profile (on hosted exchange server)
- Import PST and NK2
- Verify all email, calendars, etc. are correct
- Set Rackspace as default profile

5.4.2. Notes:

5.4.3. Additional Work / Issues.

Enter here and then copy to Final Stage

5.4. Stage Eight

Service Request # 1578

(Repeat this page as needed until all necessary stages are defined.)

5.4.1. Actions to be taken:

Point spam filter to Rackspace

NOTE: At this point, almost all email is on the new server

And all new email will go to the new server

We still need to go back and get the last little bit of email

That arrived after the PSTs were imported and before

The spam filter was re-directed

5.4.2. Notes:

5.4.3. Additional Work / Issues.

Enter here and then copy to Final Stage

5.4. Stage Nine

Service Request # 1579

(Repeat this page as needed until all necessary stages are defined.)

5.4.1. Actions to be taken:

Import final emails.

See checklist.

For each client account

Open the old profile and the new profile at the same time

(requires Outlook 2010 or 2013).

And move the last few emails to the new profile mailbox

Remove the old profile from the user's Outlook. That way they won't be confused.

5.4.2. Notes:

5.4.3. Additional Work / Issues.

Enter here and then copy to Final Stage

5.4. Stage Ten

Service Request # 1580

(Repeat this page as needed until all necessary stages are defined.)

5.4.1. Actions to be taken:

Shut down Exchange in-house

When all mail has been moved,

And all users are successfully sending and receiving

Via the new hosted email service,

Shut down and disable the Exchange services on the old

In-house Exchange server.

5.4.2. Notes:

5.4.3. Additional Work / Issues.

Enter here and then copy to Final Stage

5.5. Fine-Tuning to make sure all stages are complete and all goals accomplished

Check with client, end-users, vendors, etc. Verify that all individual stages of work are complete and that they all work well together.

5.5.1. Notes:

Mary in Sales says that some emails are missing. She may
Have created rules to move them and then forgot. Create
An SR and determine whether it's cleanup for this project
Or a separate issue.

5.5.2. Additional work or fine-tuning required.
If work is minor, define and perform it here. If it is more involved, copy to the Final Stage.

Some users did not have email on the old system. They
need to be set up. Separate SR created.

5.6. Documentation

All work should be documented as we go through the stages.

At this point, we need to verify that all documentation is complete, both in the Network Documentation Binder and in PSA system.

5.6.1. Notes:

Update our standard handouts re: OWA.

Notes entered in Network Documentation Binder along with project narrative.

+++

5.7. Additional Work Required

Use this page to define all "additional" work that comes up during the project.

Determine whether the work is **inside** or **outside** the scope of the original project.

Determine whether work will be completed now (minor work only) or placed into a separate Service Request or Project.

5.7.1. Additional Work

Work Description	Disposition / SR
Some users need to be set up.	1625

Mary Sales "lost" emails. Get mail. Inside scope. Do it.

5.7.1. Notes:

5.8. Client Training

Complete Client Training as defined in 5.2. above.

Notes:

Client: Train group on OWA.

-- Memo with how-to information

Completed July 11 *RP*

5.9. Project Completed and Signed Off

Verify with client that project is completed, all Service Requests are closed (completed).

Notes:

Agreed. Client's happy. Email's flowing. Zero downtime. *RP*

6. 0 Final check-off for project plan

Meet with the decision-making stakeholders and verify that the plan is correct, that all objectives will be met, and agree on what constitutes the scope of the project.

After this step is complete, you may begin implementing the plan.

Met July 1 with client. Agreed this is the plan. OK to proceed. *RP*

7.0. Project Evaluation

Project manager and all technicians go through the project and determine what went well, what needs improvement, etc.

Notes:

Went very smoothly.

Only hitch: One tech spent too much time exporting and importing emails instead of using Outlook 2013 to open two profiles at the same time.

Let's do internal training on this.

In fact, ZERO downtime. Client's very happy about that.

Probably need to schedule some additional billable training on maintaining mailboxes.

Section 6 - Appendices

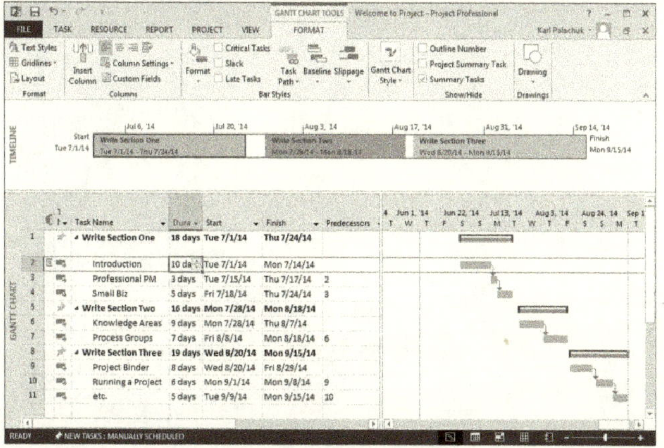

Appendix A:
The Binder Forms

The following pages contain the binder forms you will find in this kit.

Building a binder is very straight forward. Print off a set of forms, three-hole punch them, and put them in a binder.

Note that the actual forms in Microsoft Word format are included in the downloadable material for this book. See the beginning of this book for instructions on registering your book and downloading these forms.

The actual Word docs are slightly different from the forms here, simply because we had to format these for the smaller book format.

KPEnterprises

Project Binder

____

(Project Title)

____

Project Manager: ______________________

Project Leader: ______________________

Due Date: ______________________

KPEnterprises Project Binder

Status

Current Status / Last Action	Next Action Step	Date

Table of Contents

Project Description

1.0 Overview

Describe the primary goals of this project and how they will be achieved.

2.0 Define Goals for the Project

Goals should be visible, attainable actions or results. You should be able to point to something and say: "This represents the completion of the goal." So, in addition to stating each goal, also state how you will verify completion of the goal.

3.0 Define where the project (network, etc.) is today. What is the status quo?

Document current status. This may include descriptions, summaries, network maps, etc.

4. 0 Define where you want to be.

Document what the system will look like when you are finished. That is, after you have met all goals, what will it look like?

5. 0 Define all stages and steps needed to meet all objects and goals.

Major stages generally include the following:

5.1. Discovery – Documenting pre-existing network

5.2. Define training needed by client or technicians

5.3. Define the timeline

5.4. Stage One

- Actions
- Notes
- Additional work/issues to "Additional Work" stage

Note: Repeat stages as needed until all work defined. There will be as many stages as there needs to be.

5.5. Fine-Tuning to make sure all stages are complete and all goals accomplished

5.6. New system documented

5.7. Additional work stage

All "additional" work that comes up flows to this stage.

5.8. Deliver client training if needed

5.9. Project completed and signed off

Notes:

5.1 Discovery

Document the pre-existing network and all other elements relevant to this project.

5.2. Define Training Needed by Client or Technicians

5.3. Timeline

Action, Stage, or Deliverable	Due Date

5.4. Stage One ____________

Service Request # ________________

5.4.1. Actions to be taken:

5.4.2. Notes:

5.4.3. Additional Work / Issues.

Enter here and then copy to Final Stage

5.4. Stage ________________

Service Request # __________

(Repeat this page as needed until all necessary stages are defined.)

5.4.1. Actions to be taken:

5.4.2. Notes:

5.4.3. Additional Work / Issues.

Enter here and then copy to Final Stage

5.5. Fine-Tuning to make sure all stages are complete and all goals accomplished

Check with client, end-users, vendors, etc. Verify that all individual stages of work are complete and that they all work well together.

5.5.1. Notes:

5.5.2. Additional work or fine-tuning required.
If work is minor, define and perform it here. If it is more involved, copy to the Final Stage.

5.6. Documentation

All work should be documented as we go through the stages.

At this point, we need to verify that all documentation is complete, both in the Network Documentation Binder and in PSA system.

5.6.1. Notes:

5.7. Additional Work Required

Use this page to define all "additional" work that comes up during the project.

Determine whether the work is **inside** or **outside** the scope of the original project.

Determine whether work will be completed now (minor work only) or placed into a separate Service Request or Project.

5.7.1. Additional Work

Work Description	Disposition / SR
______________________________	____________
______________________________	____________
______________________________	____________
______________________________	____________
______________________________	____________

5.7.1. Notes:

5.8. Client Training

Complete Client Training as defined in 5.2. above.

Notes:

5.9. Project Completed and Signed Off

Verify with client that project is completed, all Service Requests are closed (completed).

Notes:

< This must be completed before the project begins. >

6. 0 Final check-off for project plan

Meet with the decision-making stakeholders and verify that the plan is correct, that all objectives will be met, and agree on what constitutes the scope of the project.

After this step is complete, you may begin implementing the plan.

< This must be completed after the project is completed. >

7.0. Project Evaluation

Project manager and all technicians go through the project and determine what went well, what needs improvement, etc.

Notes:

Appendix B: Definitions and Acronyms

Agile PM — Agile Project Management. A methodology in which the project work is broken down into bite-sized pieces. Each piece is performed as a whole with corrections or changes being incorporated on the fly.

CSM — Certified Scrum Master. Someone who manages the agile components in Agile project management.

DNS — Domain Name Service

FQDN — Fully Qualified Domain Name. The complete Internet name space address of a specific machine, in the format *machine.domain.com.*

LOB	Line of Business application. A type of software specific to a given industry (or line of business).
NDB	Network Documentation Binder
PM	Project Management
PMBOK	The PMBOK Guide (The Project Management Body of Knowledge)
PMI	Project Management Institute (See www.PMI.org)
PMP	Project Management Professional
Project	Any undertaking that requires more than two steps.
PSA	Professional Services Administration. A type of software that includes modules for running your professional service business.

RDNS	Reverse DNS (Reverse Domain Name Service)
ROI	Return on Investment
RWA	Remote Web Access
RWW	Remote Web Workplace
SMB	Small and Medium Business
SBS	Small Business Server
Scope	(Scope of Work) is the set of tasks that are considered to be "inside" the project. Everything else, no matter how important, is outside the project.
Scope Creep	The tendency for a project to change while in progress. This might include changing the overall goals. But more often, it simply consists of adding little tasks as you go along.

SDLC	Software Development Life Cycle
Service Request	A distinct task that needs to be completed. In a service tracking system, each service request has a unique number or other identifier so you can keep track of all work related to that task.
Service Ticket	See Service Request
SMB	Small and Medium Business.
SOW	Statement of Work (see also Scope or Scope of Work)
SRV Record	A Service Record for DNS, for example to define that an external email server may legitimately send on behalf of your domain.
Stakeholder	Any individual, organization, or group who may affect, be affected by, or perceive itself to be affected by a decision, activity, or outcome of a project.

Appendix C: Resources

Books

The E-Myth Revisited by Michael Gerber

Getting Things Done by David Allen

Network Documentation Workbook by Karl W. Palachuk

The PMBOK Guide – 5th ed. The Project Management Body of Knowledge

Software and Services

Autotask. See www.autotask.com.

ConnectWise PSA. See www.connectwise.com.

Hosted Spam Filter. There are many services available two examples are www.reflexion.net and www.exchangedefender.com.

Project+ exam from Comptia. See www.comptia.org.

Project Management Institute. See www.PMI.org.

Templates for Project Management

(see discussion in Chapter Four. Note: Most universities and government IT branches offer free stuff on-line.)

Businessballs.com - http://www.businessballs.com/project%20management%20templates.pdf

Excel stuff - http://www.makeuseof.com/tag/excel-project-management-tracking-templates/

VITA (Virginia Information Technologies Agency) - http://vita.virginia.gov/oversight/projects/default.aspx?id=567

UC Santa Cruz - http://its.ucsc.edu/project-management/templates.html#Manage-a-project

North Dakota IT - http://www.nd.gov/itd/standards/project-management/project-management-guidebooks-and-templates

Other Resources from Small Biz Thoughts

Please Check Out Our Web Sites:

www.SMBBooks.com

This is our primary site for books on technical topics, managed services, running your business, and more. All of our up-coming training events and recorded programs are there as well.

www.SmallBizThoughts.com

blog.SmallBizThoughts.com

This is our primary web site and Karl's popular blog for I.T. Consultants and Managed Service Providers. You can also find out about SOPs (standard operating procedures) and business coaching through this web site.

Karl's Weekly Newsletter

Register at one of the sites above or at GreatLittleBook.com.

This newsletter covers upcoming events, seminars, news, and "what's happening" in the SMB Consulting space.

Please also consider these fine books by Great Little Book:

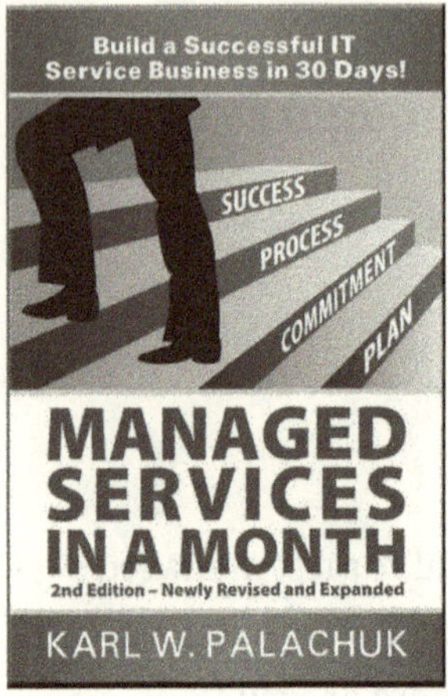

Managed Services in a Month 2nd ed.

Build a Successful IT Services Business in 30 Days.

by Karl W. Palachuk

2013

208 pages

A no-nonsense guide to building a successful managed service practice.

Whether you are just starting out, or converting your existing break/fix technology consulting business to managed services, this book will show you the way. The newly revised and expanded 2nd edition has nine new chapters, covering the latest products and services available today-including cloud technologies.

Also available as an e-book, audio book, or in a Spanish language translation.

The #1 book on Managed Services on Amazon.com for more than five years!

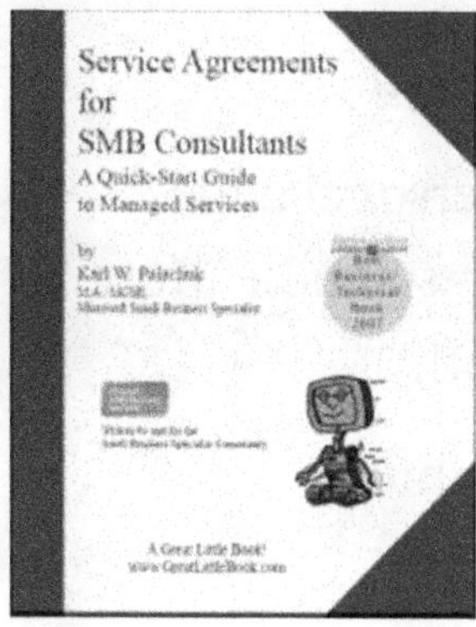

Service Agreements for SMB Consultants

A Quick-Start Guide to Managed Services

by Karl W. Palachuk

2006

185 pages

This great little book does a lot more than give you sample agreements.

Karl starts out with a discussion of how you run your business and the kinds of clients you want to have. The combination of these – defining yourself and defining your clients – is the basis for your service agreements.

Includes sample contracts with commentaries. All text, as well as some other great resources are provided as downloads.

Available in paperback or e-book formats.

---

www.SMBBooks.com

www.SmallBizThoughts.com

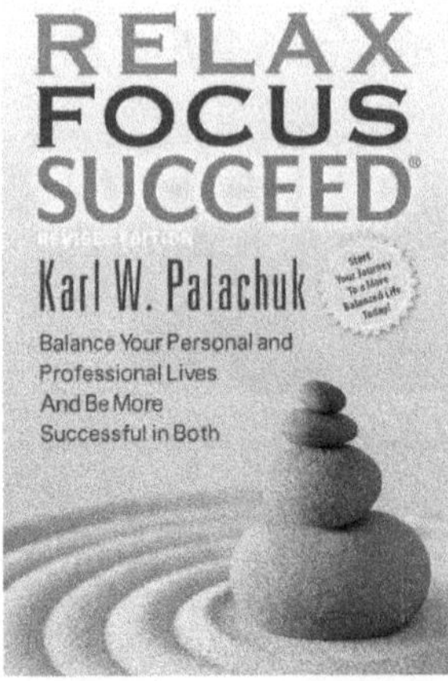

Relax Focus Succeed®

Balance Your Personal and Professional Lives and Be More Successful in Both

by Karl W. Palachuk

2007

296 pages

The premise of this book is simple but powerful: The fundamental keys to success are focus, hard work, and balance. Too often, the advice we receive gives plenty of attention to focus and hard work, but very little to balance.

This great little book will help you believe that you need balance, show you the power of focus, and help you move forward with the new you -- a happier, healthier, better balanced, and more successful you.

www.SMBBooks.com

www.SmallBizThoughts.com

www.ingramcontent.com/pod-product-compliance
Lightning Source LLC
Chambersburg PA
CBHW021031210326
41598CB00016B/985

9780976376088